THE PROBATIONER'S HANDBOOK

A MANUAL OF INSTRUCTION FOR THE STUDENT OF THE A.'.A.'.

GEORGE T. MORTIMER

A MEDIA UNDERGROUND PUBLICATION

ISBN 978-1-4092-4716-6

In this book it is spoken of the Sephiroth, and the Paths, of Spirits and Conjurations; of Gods, Spheres, Planes and many other things which may or may not exist.

It is immaterial whether they exist or not. By doing certain things certain results follow; students are most earnestly warned against attributing objective reality or philosophic validity to any of them.

(Crowley, A; *Liber O*)

Author's Note

Terms like "he", "his", or "him" have been employed throughout this work as a means of expressing both sexes. One therefore trusts that one's female readers will accept this unfortunate sexist phraseology as a means of avoiding awkward expression.

As the title suggests this work is primarily a handbook, therefore its successful application depends entirely on utilising it in accordance with other authors' work. Anyone who relies heavily on the words of the present writer alone will be in grave danger of becoming psychologically unbalanced.

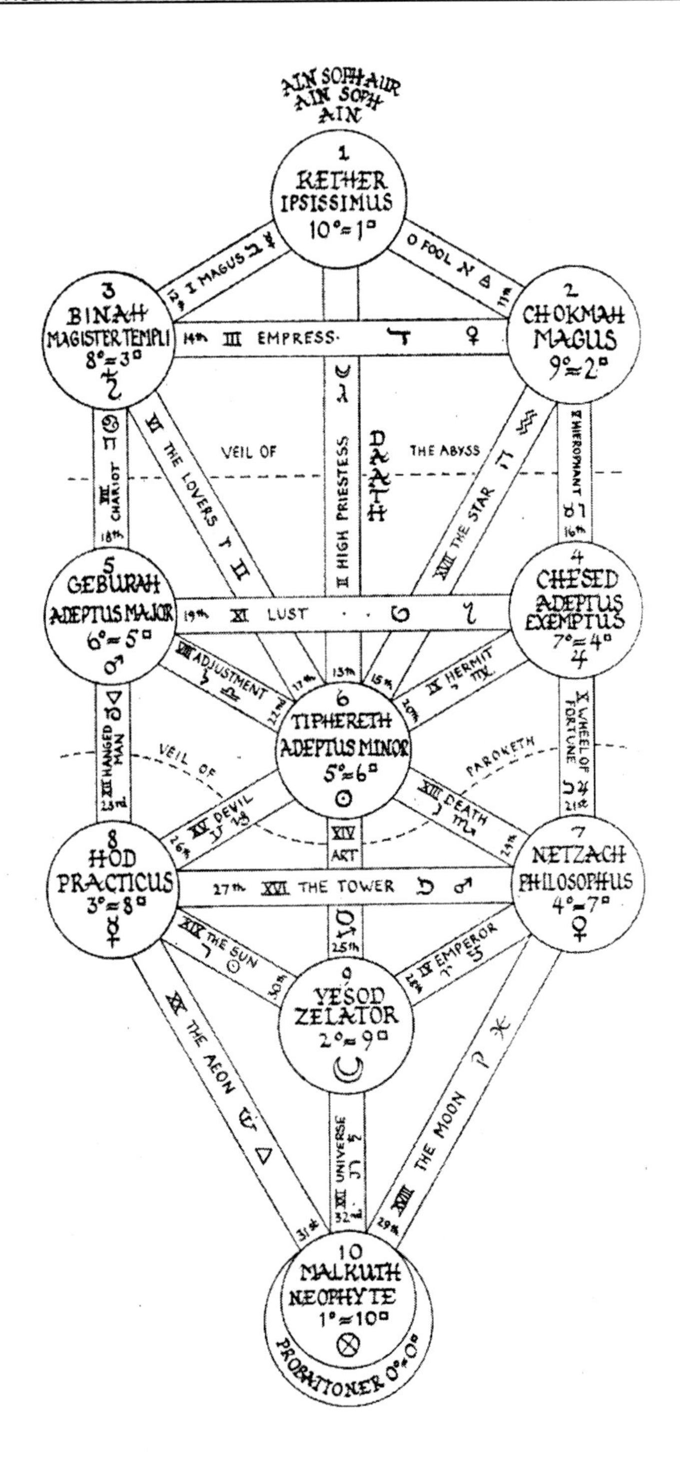
AIN SOPH AUR
AIN SOPH
AIN
1
KETHER
IPSISSIMUS
10°=1□
3
BINAH
MAGISTER TEMPLI
8°=3□
2
CHOKMAH
MAGUS
9°=2□
12th I MAGUS
O FOOL
11th
14th III EMPRESS
VEIL OF
THE ABYSS
DAATH
II HIGH PRIESTESS
13th
VI THE LOVERS
17th
VII CHARIOT
18th
XVII THE STAR
15th
V HIEROPHANT
16th
5
GEBURAH
ADEPTUS MAJOR
6°=5□
4
CHESED
ADEPTUS
EXEMPTUS
7°=4□
19th XI LUST
VIII ADJUSTMENT
22nd
IX HERMIT
20th
6
TIPHERETH
ADEPTUS MINOR
5°=6□
XII HANGED MAN
23rd
X WHEEL OF FORTUNE
21st
VEIL OF
PAROKETH
XV DEVIL
26th
XIII DEATH
24th
XIV
ART
8
HOD
PRACTICUS
3°=8□
7
NETZACH
PHILOSOPHUS
4°=7□
27th XVI THE TOWER
XIX THE SUN
30th
25th
IV EMPEROR
28th
9
YESOD
ZELATOR
2°=9□
XX THE AEON
31st
XXI UNIVERSE
32nd
XVIII THE MOON
29th
10
MALKUTH
NEOPHYTE
1°=10□
PROBATIONER 0°=0□

In memory of Gerald Suster (1951-2001)

CONTENTS

To All True Seekers Of Wisdom

PREFACE

POSTCARDS TO PROBATIONERS

by

Aleister Crowley

(Taken from *The Equinox* Volume II, pp. 196 - 200)

THEOREMS

I. The world progresses by virtue of the appearance of Christs (geniuses).
II. Christs (geniuses) are men with super-consciousness of the highest order.
III. Super-consciousness of the highest order is obtainable by known methods.

ESSENTIALS OF METHOD

I. Theology is immaterial; for both Buddha and St. Ignatius were Christs.
II. Morality is immaterial; for both Socrates and Mohammed were Christs.
III. Super-consciousness is a natural phenomenon; its conditions are therefore to be sought rather in the acts than the words of those who attain it.
The essential acts are retirement and concentration - as taught by Yoga and Ceremonial Magic.

MISTAKES OF MYSTICS

I. Since truth is supra-rational, it is incommunicable in the language of reason.
II. Hence all mystics have written nonsense, and what sense they have written is so far untrue.
III. Yet as a still lake yields a truer reflection of the sun than a torrent, he whose mind is best balanced will, if he become a mystic, become the best mystic.

THE METHOD OF EQUILIBRIUM

I. THE PASSIONS, etc

I. Since the ultimate truth of teleology is unknown, all codes of morality are arbitrary.
II. Therefore, the student has no concern with ethics as such.
III. He is consequently free "to do his duty in that state of life to which it has pleased God to call him."

II. THE REASON

I. Since truth is supra-rational, any rational statement is false.
II. Let the student then contradict every proposition that presents itself to him.
III. Rational ideas being thus expelled from the mind, there is room for the apprehension of spiritual truth.
It should be remarked that this does not destroy the validity of reasonings on their own plane.

III. THE SPIRITUAL SENSORIUM

I. Man being a finite being, he is incapable of apprehending the infinite. Nor does his communion with infinite being (true or false) alter this fact.

II. Let then the student contradict every vision and refuse to enjoy it; first, because there is certainly another vision possible of a precisely contradictory nature; secondly, because though he is God, he is also a man upon an insignificant planet.

III. Being thus equilibrated laterally and vertically, it may be that, either by affirmation or denial of all these things together, he may attain the supreme trance.

IV. THE RESULT

I. Trance is defined as the ek-stasis of one particular tract of the brain, caused by meditating on the idea corresponding to it.

II. Let the student therefore beware lest in that idea be any trace of imperfection. It should be pure, balanced, calm, complete, fitted in every way to dominate the mind, as it will.

Even as in the choice of a king to be crowned.

III. So will the decrees of this king be just and wise as he was just and wise before he was made king.

The life and works of the mystic will reflect (though dimly) the supreme guiding forces of the mystic, the highest trance to which he has attained.

YOGA AND MAGIC

I. Yoga is the art of uniting the mind to a single idea.

It has four methods.

	Gnana-Yoga.	Union by Knowledge.
	Raja-Yoga.	Union by Will.
	Bhakta-Yoga.	Union by Love.
add	Mantra-Yoga.	Union through Speech.
	Karma-Yoga.	Union through Work.

These are united by the supreme method of Silence.

II. Ceremonial Magic is the art of uniting the mind to a single idea.

It has four methods.

	The Holy Qabalah.	Union by Knowledge.
	The Sacred Magic.	Union by Will.
	The Acts of Worship.	Union by Love.
	The Ordeals.	Union by Courage.
add	The Invocations.	Union through Speech.
	The Acts of Service.	Union through Work.

These are united by the supreme method of Silence.

III. If this idea be any but the Supreme and Perfect idea, and the student lose control, the result is insanity, obsession, fanaticism, or paralysis and death (add addiction to gossip and incurable idleness), according to the nature of the failure.

Let then the Student understand all these things and combine them in his Art, uniting them by the supreme method of Silence.

-oOo-

FOREWORD

Each year, practically hundreds of utterly fatuous books are written on the subject of Occultism. The outcome of such nonsense *does not*, as on imprudent writer put it, "[allow us] to see how universal principles, symbols and myths are outer representations of a deep collective psycho-mythology that is working inherently throughout the human species"[1]. What such books do achieve is an ambience of total irrelevance, whereby one particularly appalling writer makes a hell of a lot of cash, whilst at the same time degrading the occult scene into a collection of babbling idiots.

Fortunately, however, there is a small percentage of gold amongst the dross. One need only study the works of Israel Regardie, Francis King, Robert Anton Wilson or Gerald Suster to discern the truth in this statement. Aleister Crowley was also such gold; a former initiate of the Hermetic Order of the Golden Dawn, he became one of the first individuals to truly revolutionise the magical arts. He was also the pioneer of what he aptly called "Scientific Illuminism" - a means of attaining heightened states of consciousness by employing scientific methods.

The term "Scientific Illuminism" was first coined in 1909, in a beautifully produced biannual journal entitled *The Equinox* (so-called because of its appearance at the Spring and Autumnal Equinoxes). Subtitled "The Review of Scientific Illuminism", *The Equinox* bore two curious maxims upon its front cover: "The Aim of Religion" and "The Method of Science".

Crowley, an established authority on occult matters, believed in the existence of "praeter-human intelligence," and that our evolutionary process could be accelerated by means of conversing with such entities[2]. He also believed that he himself had entered into such communication in 1904; the result of which was the reception of an intriguing book entitled *The Book Of The Law* or *Liber AL vel Legis*[3].

The Book Of The Law is an exceptionally curious document indeed; however, rather than embark upon a lengthy summary of its principles and message, one trusts that the interested reader will pursue the matter further in *The Equinox Of The Gods*[4] and other such works more suited to its study. Suffice it to say here then, that from a political and ethical standpoint *The Book Of The Law* advocates a form of individualistic libertarianism, suitable only for those with the most inquiring of minds.

---

[1] These ghastly words were in fact written by Angeles Arrien in her appallingly uninspiring manual, *The Tarot Handbook - Practical Applications of Ancient Visual Symbols* (Diamond Books, 1995). Here, Ms Arrien managed to write a dreadful book about Crowley's *Thoth Tarot*, without once mentioning the importance of the Qabalah (note: Diamond Books have a habit of churning out garbage; in particular, the tedious trilogy composed by Amado Crowley - Crowley's alleged son).

[2] The definition of "praeter-human intelligence" is indeed a rather difficult expression to properly clarify. Some occultists have come to understand it as meaning "one's subconscious self" whereas others regard it as an alternative expression for "Angel", "God", "extra-terrestrial", or even possibly "Devil". Throughout his life, Crowley seems to have remained relatively open to all of these interpretations.

[3] The communicating entity revealed itself as "Aiwass the minister of Hoor-paar-kraat" and actually referred to this curious document as "The Book of the Law" in the final and closing verse. This manuscript is often quite disturbing in context and announces the beginning of a new era for mankind in which the formula of death and redemption (the formula of Christ, the Aeon of Osiris) is replaced rather violently by a formula of equality and individuality (the Aeon of Horus).

[4] Crowley, A; New Falcon Publications, 1991.

In 1907, Aleister Crowley and another former Golden Dawn initiate, George Cecil Jones[5] founded the Magical Order of the A∴A∴[6], an occult organisation dedicated to the evolution of human consciousness. By 1909 the A∴A∴ had embraced the doctrines of *The Book Of The Law*, and *The Equinox* was duly ascribed as the Order's "Official Organ". One must remark that within the pages of the aforementioned journal, one finds the most exceptional form of magical instruction, which, in the present writer's opinion, has still to this day to be paralleled. As Crowley remarked in his autohagiography[7]:

> *The Equinox* was the first serious attempt to put before the public the facts of occult science... From the moment of its appearance, it imposed its standards of sincerity, scholarship, scientific seriousness and aristocracy of all kinds... It is recognised as the standard publication of its kind, as an encyclopaedia without "equal, son, or companion". It has been quoted, copied, and imitated everywhere... Its influence has changed the whole current of thought of students all over the world.[8]

Let the student therefore commence study of Occultism with equal exuberance; employing such standards of sincerity, scholarship, scientific seriousness and aristocracy wherever at all possible. It must be clearly understood that Magick is by no means some foolhardy pursuit. If one is to get anywhere and achieve anything, laziness and inactivity must not be allowed to prevail. This is a path which undoubtedly requires a considerable amount of self-discipline, for as the late Dr. Israel Regardie once put it:

> Nothing in the world can take the place of persistence.
>
> Talent will not; nothing is more common than unsuccessful men with talent.
>
> Genius will not; unrewarded genius is almost a proverb.
>
> Education will not; the world is full of educated derelicts.
>
> Persistence and determination alone are omnipotent.[9]

-oOo-

---

[5] Crowley described Jones as "a Welshmen" who "possessed a fiery but stable temper, was the son of a suicide, and bore a striking resemblance to many conventional representations of Jesus Christ. His spirit was both ardent and subtle. He was very widely read in Magick; and, being by profession an analytical chemist, was able to investigate the subject in a scientific spirit."

[6] In *The Magick Of Thelema* (Samuel Weiser, Inc, 1993), Lon Milo DuQuette remarks that "It is commonly believed that A∴A∴ stands for Argenteum Astrum (Silver Star). I have been informed in no uncertain terms that this is not the case". Yet, rather than actually shed some light on this intriguing matter, Mr. DuQuette has, like so many of his predecessors, decided to keep us in the dark with regards to the truth. Unfortunately, the present writer has no idea what Mr. DuQuette is talking about. If A∴A∴ does indeed stand for something else, then one doubts very much if it matters a damn. It is after all the techniques of the Order which are important to us, not some closely guarded secret of meritless worth.

[7] Autohagiography is a term that was employed by Crowley to mean "the autobiography of a saint".

[8] Crowley, A; *The Confessions of Aleister Crowley*, page 604 (Arkana Books, 1989).

[9] Regardie, I; *The Complete Golden Dawn System Of Magic* (New Falcon Publications, 1990).

INTRODUCTION

There are indeed many fine works for the occult student to scrutinise. Israel Regardie's *The One Year Manual*[10] is perhaps one of the finest publications currently available for those wishing to embark upon a year's self-discovery. However, although Dr. Regardie expresses the notion that his work might be of value to the A∴A∴ Probationer, the present writer does feel that the book lacks a certain amount of substructure concerning the Probationary tasks. It is for that reason alone, that one finds Regardie's *Gems From The Equinox*[11] of far greater assistance. However, *Gems* does also cater for those of a much higher magical grade; and this, one feels, might bewilder the Probationer slightly with regards to some of the more advanced techniques mentioned therein.

Another splendid introductory work currently available on the market is Gerald del Campo's *New Aeon Magick*[12]. Here, del Campo presents a delightful prelude to the rudiments of the Thelemic path. Unfortunately he ruins an otherwise excellent book by insisting that the diligent student should seek instruction from a chartered A∴A∴ Order. As he misguidedly points out: "Anyone claiming membership in this Order should be able to provide a clean, unbroken lineage beginning with Aleister Crowley. This information should be provided verbally, or via some documentation". One doubts very much if there are many such groups still in existence. Furthermore, as the provocative occult writer Gerald Suster once told me: "If a group is doing good magical work, it doesn't need a charter. If it isn't, then a charter won't save it". Therefore, one strongly recommends that one steers well clear of groups offering instruction in return for large sums of money. As Crowley himself remarked: "There is... an absolute prohibition to accept money or other material reward, directly or indirectly, in respect of any service connected with the Order, for personal profit or advantage. The penalty is immediate expulsion, with no possibility of reinstatement on any terms."[13]

It is indeed a largely believed falsehood that one requires affiliation in order to attain the grades of the A∴A∴ syllabus. If one is sincere enough and hard working, the necessary magical link can be established relatively easily. One will find that the Universe has a tendency to push one forward using one's own generated momentum. The present writer not only recommends this course of action, but would also like to advise the student against becoming involved with anyone boasting of Adeptship. Such people encompass a large section of the esoteric community and are solely responsible for giving the occult scene a bad name. They tend to be obsessive megalomaniacs, exclusively concerned with their own self-importance and magical attainments. It is therefore extremely consequential for the student to get into the habit of never uttering his magical grade to anyone out of pomposity[14]. Firstly, because it is the business of nobody else; secondly, because you are not that bloody important. As you steadily advance through the grades of the system (and, one may add, through the

[10] Samuel Weiser, Inc, 1990.
[11] Falcon Press, 1986.
[12] Llewellyn Publications, 1994.
[13] Crowley, A; *One Star In Sight*.
[14] In fact there is no need to utter it at all.

grades of life), such matters will become increasingly more self-evident. As it is written in the document *Liber Librae*: "He who knoweth little, thinketh he knoweth much; but he who knoweth much hath learned his own ignorance." There are far too many casualties on the occult scene already without striving to add further to the list[15].

My intentions then, for composing this little work can be thus considered threefold:

1. To make available a manual of instruction suitable for those either currently working through the Probationary tasks, or wishing to begin them.
2. To fulfil an oath of my own taking and "observe zeal in service of the Probationers under me."[16]
3. To revise and update a very pure, noble and beautiful system that is all too often ignored by so-called magicians working within a Thelemic context.

One therefore trusts that the fundamentalists will forgive me for my innovations and alterations to some of Uncle Aleister's prescribed techniques. As Crowley himself remarked: "the student... if he have any capacity whatever... will find his own crude rituals more effective than the highly polished ones of other people."[17] This is good advice for the student to commence with. Do not do something the nature of which you do not understand. Study the matter thoroughly, digest all the information given, assimilate it into you own way of thinking, and if necessary make alterations. If you cannot learn to do this, you will not make much of a magician.

Go to it then, and may your quest be rewarding and sublime!

-oOo-

[15] As I sit here today (15th January 2007) and rewrite this book to bring it up to date, I am even more astounded by the amount of casualties I have witnessed. Many of whom I considered were far too intelligent and careful to be drawn into the Pit of Desolation.

[16] Taken from the Oath of a Neophyte.

[17] *Magick*, page 449, *Liber O vel Manus et Sagittae* (Guild Publishing, 1989).

CHAPTER ONE

COMMENCING THE WORK
(SOME PRELIMINARY REMARKS FOR THE BENEFIT OF THE STUDENT)

Before committing oneself to any work, it is of considerable importance for the student to acquire a basic knowledge of what he or she is embarking upon. It is for that reason then that the present writer has included excerpts from Crowley's essay on the matter, *One Star In Sight*:

A∴A∴

1. The Order of the Star called S. S. is, in respect of its existence upon the Earth, an organised body of men and women distinguished among their fellows by the qualities here enumerated. They exist in their own Truth, which is both universal and unique. They move in accordance with their own Wills, which are each unique, yet coherent with the universal will.
   They perceive (that is, understand, know, and feel) in love, which is both unique and universal.

2. The order consists of eleven grades or degrees, and is numbered as follows: these compose three groups, the Orders of the S. S., of the R. C., and of the G. D.[18] respectively.

The Order of the S. S.

Ipsissimus	10° = 1 ▫
Magus	9 ° = 2 ▫
Magister Templi	8 ° = 3 ▫

The Order of the R. C.
(Babe of the Abyss - The Link)

Adeptus Exemptus	7 ° = 4 ▫
Adeptus Major	6 ° = 5 ▫
Adeptus Minor	5 ° = 6 ▫

The Order of the G. D.
(Dominus Liminis - The Link)

Philosophus	4 ° = 7 ▫
Practicus	3 ° = 8 ▫
Zelator	2 ° = 9 ▫
Neophyte	1 ° = 10 ▫
Probationer	0 ° = 0 ▫

The general attributions of these Grades are indicated by their correspondences on the Tree of Life (as shown on page 6 of this manual).

Student. - His business is to acquire a general intellectual knowledge of all systems of attainment, as declared in the prescribed books.
Probationer. - His principal business is to begin such practices as he may prefer, and to write a careful record of the same for one year.

[18] The Silver Star, the Rosy Cross and the Golden Dawn.

Neophyte. - Has to acquire perfect control of the Astral Plane.
Zelator. - His main work is to achieve complete success in Asana and Pranayama. He also begins to study the formula of the Rosy Cross.
Practicus. - Is expected to complete his intellectual training, and in particular to study the Qabalah.
Philosophus. - Is expected to complete his moral training. He is tested in Devotion to the Order.
Dominus Liminis. - Is expected to show mastery of Pratyahara and Dharana.
Adeptus (without). - Is expected to perform the Great Work and to attain the Knowledge and Conversation of the Holy Guardian Angel.
Adeptus (within). - Is admitted to the practice of the formula of the Rosy Cross on entering the College of the Holy Ghost.
Adeptus (Major). - Obtains a general mastery of practical Magick, though without comprehension.
Adeptus (Exemptus). - Completes in perfection all these matters. He then either (a) becomes a Brother of the Left Hand Path or, (b) is stripped of all his attainments and of himself as well, even of his Holy Guardian Angel, and becomes a Babe of the Abyss, who, having transcended the Reason, does nothing but grow in the womb of its mother. It then finds itself a Magister Templi.
Magister Templi. - (Master of the Temple): whose functions are fully described in *Liber 418*, as is this whole initiation from Adeptus Exemptus. See also *Aha!* His principal business is to tend his "garden" of disciples, and to obtain a perfect understanding of the Universe. He is a Master of Samadhi.
Magus. - Attains to wisdom, declares his law and is a Master of all Magick in its greatest and highest sense.
Ipsissimus. - Is beyond all this and beyond all comprehension of those of lower degrees.

This then is a brief description of the path set forth before the student. The reader will observe that his first task is to "acquire a general intellectual knowledge of all systems of attainment". This can be achieved by studying the relevant works set forth in Appendix I of Crowley's *Magick*. However, although pursuit of such a reading curriculum would be of undoubted value to the student, the present writer does consider it to be rather unnecessarily lengthy. As Israel Regardie remarked in his Preface to *The One Year Manual*:

> The occult student, at the onset of his studies, is besieged by hundreds of books describing dozens of practices of every kind. They promise, directly or otherwise, to bring him to the very heights of spiritual attainment, no matter how that attainment is defined. By the very wealth of material is he overwhelmed. And the result is that, generally speaking, he does nothing except read. Reading does very little to bring one to any kind of realisation of one's divine nature.

The student then should keep in mind the underlying intention for pursuing such a reading curriculum in the first place, which is, as Crowley pointed out:

> ....to familiarise the student with all that has been said by the Great Masters in every time and country. He should make a critical examination of them; not so much with the idea of discovering where truth lies, for he cannot do this except by virtue of his own spiritual experience, but rather to discover the essential harmony in these varied works. He should be on his guard against partisanship with a favourite author. He should familiarise himself thoroughly with the method of mental equilibrium, endeavouring to contradict any statement soever, although it may be apparently axiomatic.
>
> The general object of this course, besides that already stated, is to assure sound education in occult matters, so that when spiritual illumination comes it may find a well-built temple. Where the mind is strongly biased towards a special theory, the result of an illumination is often to inflame that portion of the mind which is thus overdeveloped, with the result that the aspirant, instead of becoming an Adept, becomes a bigot and a fanatic.[19]

[19] *Magick*, Appendix I.

Therefore it seems reasonable to assume that one could perhaps commence study of a somewhat less demanding reading list, providing of course that one maintains one's level of "mental equilibrium". The following list then, first appeared in Crowley's *Liber E vel Exercitiorum*, and is undoubtedly a much more realistic approach. The present writer hopes that these works will form a basis of the student's library, for they are all invaluable manuals and give considerable insight into the nature of the Great Work.

The Yi King.
The Tao The King.
Tannhäuser by A. Crowley.
The Upanishads.
The Bhagavad-Gita.
The Voice Of The Silence by H.P. Blavatsky.
Raja Yoga by Swami Vivekananda.
The Shiva Sanhita.
The Aphorisms Of Patanjali.
The Sword Of Song by A. Crowley.
The Book Of The Dead.
Dogme Et Rituel De La Haute Magie by E. Levi.
The Book Of The Sacred Magic Of Abramelin The Mage.
The Goetia.
The Hathayoga Pradipika.
The Spiritual Guide Of Molinos.
A History Of Philosophy by J.E. Erdmann.
The Star In The West by J.F.C. Fuller.
The Dhammapada.
The Questions Of King Milinda.
777 Vel Prolegomena, etc by A. Crowley and others.
Varieties Of Religious Experience by W. James.
Kabbala Denudata by S.L. MacGregor Mathers.
Knox Om Pax by A. Crowley.[20]

Some of these works might be difficult to procure these days,[21] therefore rather than have the student search endlessly for titles that are no longer available in print, he might gain further insight into the path set before him by additionally studying the following:

[20] Looking at this list as I update this book today (15th January 2007), I find the over emphasis placed upon eastern philosophy to be somewhat irritating. Indeed it is true that much wisdom has come from the east, yet equally there has been just as much wisdom generated in the west. The reader will no doubt judge best for himself, however at the conclusion of my first writing of this book I was accused of not having included enough material on eastern methods of spiritual attainment. Partially this is due to my own personal preferences, since my nature is more active than passive; however judging by the amount of eastern hogwash that is now available in the 21st century (i.e. your average housewife's yoga video collection), I am glad that I made efforts at the time to readdress the balance.

The small amount of eastern techniques that I have included in this book are technically sound since the essence of Yogic practice, as Crowley correctly stated, is to learn to "Sit still. Stop thinking. Shut up. Get out". You could immerse yourself, from now until the end of time, in as much eastern guff as you want, and will get absolutely nowhere unless you learn to filter out the worthy techniques from the dross.

[21] Oh what an easy life the kids of today have (2007). No more hunting through second hand bookshops and getting queer looks from dusty old tarts working behind the counter. Just go online; all these manuscripts are probably available electronically in cyberspace.

Works by Aleister Crowley
Magick, Book Four, Liber ABA.
The Law Is For All.
Eight Lectures On Yoga.

Works on Aleister Crowley & Related Subjects
Gems From The Equinox by Israel Regardie.
The Legacy Of The Beast by Gerald Suster.
The Magick Of Thelema by Lon Milo DuQuette.
New Aeon Magick by Gerald del Campo.

Other Additional Works of Interest
The Golden Dawn by Israel Regardie.
The One Year Manual by Israel Regardie.
The Truth About The Tarot by Gerald Suster.
Hitler: Black Magician by Gerald Suster.
The Golden Bough by J.G. Frazer.
Tantra: The Way Of Action by Francis King.
The Doors Of Perception by Aldous Huxley.
Magic: An Occult Primer by David Conway.
The Outsider by Colin Wilson.
The Cosmic Trigger by Robert Anton Wilson.
Prometheus Rising by Robert Anton Wilson.
Rebels & Devils by various (edited by Christopher S. Hyatt).
Chaos & Cyber Culture by Timothy Leary.
The Key Of It All (AL=L) by George T. Mortimer.

Once one has acquired a basic theoretical knowledge and understanding of practical occultism - whereby one accepts the Law of Thelema as a natural and logical progression - one is then ready to proceed to the next preliminary level: that of a Probationer.

-oOo-

CHAPTER TWO

COMMENCING THE WORK
(THE TASK OF A PROBATIONER)

0. Let any person be received by a Neophyte, the latter being responsible for his Zelator.
1. The period of Probation shall be at least one year.
2. The aspirant to the A.'.A.'. shall hear the Lection (*Liber LXI*) and this note of his office; IF HE WILL, shall then procure the robe of a Probationer; shall choose with deep forethought and intense solemnity a motto.
3. On reception, he shall receive the robe, sign the form provided and repeat the oath as appointed, and receive the First Volume of the Book.
4. He shall commit a chapter of *Liber LXV* to memory; and furthermore, he shall study the Publications of the A.'.A.'.in Class B, and apply himself to such practices of Scientific Illuminism as seemeth him good.
5. Besides all this, he shall perform any tasks that the A.'.A.'. may see fit to lay upon him. Let him be mindful that the word Probationer is no idle term, but that the Brothers will in many a subtle way *prove* him, when he knoweth it not.
6. When the sun shall next enter the sign under which he hath been received, his initiation may be granted unto him. He shall keep himself free from all other engagements for one whole week from that date.
7. He may at any moment withdraw from his association with the A.'.A.'., simply notifying the Neophyte who introduced him.
8. He shall everywhere proclaim openly his connection with the A.'.A.'. and speak of It and Its principles (even so little as he understandeth) for that mystery is the enemy of Truth. One month before the completion of his year, he shall deliver a copy of the Record of his year's work to the Neophyte introducing, and repeat to him his chosen chapter of *Liber LXV*.
9. He shall hold himself chaste, and reverent towards his body, for that the ordeal of initiation is no light one. This is of peculiar importance in the last two months of his Probation.
10. Thus and not otherwise may he attain the great reward: YEA, MAY HE OBTAIN THE GREAT REWARD!

These are the official tasks as set forth in the document *Liber Collegii Sancti*[22]. However, having been written some time ago, some parts of the curriculum have become somewhat anachronistic due to human evolution and the development of new magical innovations; therefore it is perhaps of value to include a brief commentary on each of the eleven points:

0. Let any person be received by a Neophyte, the latter being responsible for his Zelator.

In essence, any person whatsoever may join the order simply by establishing contact with one of its members. It must be remarked, however, that initially, upon the Order's founding, the A.'.A.'. was meant to function as a continuous chain of grades (that is to say that the system was supposed to work on a supervisory basis) where a Practicus, for example, would be responsible for the progress of his Zelator; therefore the Zelator would in turn be responsible for his Neophyte; and the Neophyte would hold guardianship over his Probationer. Needless to say, this system never quite operates successfully. All too often the links of the chain become broken, and the Order ceases to operate effectively. This, of course, does not mean that the Order's existence is terminated (as dubiously supposed by some scholars). As will be shown

[22] First published in the Appendix of Regardie's *Gems From The Equinox*.

in the following chapter, the magical link can still be established fittingly by the sincerity of the aspirant alone.

> 1. The period of Probation shall be at least one year.

Nevertheless, there is no race to complete the period of Probation, if one wishes to take longer than a year then do so. This then is the great beauty of working the system on one's own; where one does not feel obliged to advance to the next stage simply because one's Neophyte is ready to.

> 2. The aspirant to the A∴A∴ shall hear the Lection (*Liber LXI*) and this note of his office; IF HE WILL, shall then procure the robe of a Probationer; shall choose with deep forethought and intense solemnity a motto.

The Lection (*Liber LXI*) is a History Lection giving Crowley's version of the origins of the Golden Dawn and therefore the origins of the A∴A∴'s predecessors. Due to its unnecessary complexity it is not reproduced within the confines of this work, however a modern interpretation is given at the end of this commentary.

It is of considerable importance for the aspirant to choose solemnly a motto, for since he is embarking upon a new way of life, he must also choose a new name which represents the heights of his aspiration.

> 3. On reception, he shall receive the robe, sign the form provided and repeat the oath as appointed, and receive the First Volume of the Book.

The robe can be easily made or purchased and should be black in the shape of a T. The reason for it being this colour is entirely symbolic and does not therefore possess any sinister connotations. Black is the only colour which absorbs and attracts all light, and is thus especially relevant in terms of the Probationer's aspirations. As Crowley himself remarked: "The Robe is that which conceals, and protects the Magician from the elements; it is the silence and secrecy with which he works, the hiding of himself in the occult life of Magick and Meditation."[23]

"The Book" is undoubtedly a reference to Crowley's *Thelema* which was an earlier version of the Caliphate OTO's compilation *The Holy Books Of Thelema*. The First Volume therefore is the document *Liber LXV*, or as it is otherwise known *The Book Of The Heart Girt With A Serpent*.

> 4. He shall commit a chapter of *Liber LXV* to memory; and furthermore, he shall study the Publications of the A∴A∴ in Class B, and apply himself to such practices of Scientific Illuminism as seemeth him good.

It is the present writer's opinion that *Liber LXV* need not be fully committed to memory. Firstly, because one could spend the time more productively memorising various important magical techniques and formulae; secondly because, as will be shown in the following chapter, the A∴A∴ (or Great White Brotherhood) is far more universal than the hierarchical grade system constructed by Crowley and Jones. One is advised, however, to at least familiarise oneself with the work in question, thereby grasping the essence of what it represents.

[23] *Magick*, part II, page 108 (Guild Publishing, 1989).

The Class B publications are as follows:
Liber VI: Liber O vel Manus et Sagittae.
Liber IX: Liber E vel Exercitiorum.
Liber XXI: Khing Kang King.
Liber XXX: Liber Librae.
Liber LVIII (On The Qabalah).
Liber LXI: Liber Causae.
Liber LXIV: Liber Israfel.
Liber LXXVII: The Book Of Thoth.
Liber LXXXIV: vel Chanokh.
Liber XCVI: Liber Gaias.
Liber D: Liber Sepher Sephiroth.
Liber DXXXVI: Liber Batrachophrenoboocosmomachia.
Liber DCCLXXVII: Book 777.
Liber DCCCLXVIII: Liber Viarum Viae.
Liber CMXIII: Liber Via Memoriae.
It is vital for one to keep a diary of one's practices, for how otherwise might one gauge one's results scientifically?

> 5. Besides all this, he shall perform any tasks that the A.·.A.·. may see fit to lay upon him. Let him be mindful that the word Probationer is no idle term, but that the Brothers will in many a subtle way *prove* him, when he knoweth it not.

This is quite straightforward and curiously applies even to those who have sworn The Oath by solitary means.

> 6. When the sun shall next enter the sign under which he hath been received, his initiation may be granted unto him. He shall keep himself free from all other engagements for one whole week from that date.

At the point when the sun again enters the sign under which The Oath was sworn, the Probationer is free to advance to the grade of Neophyte when he so chooses.

> 7. He may at any moment withdraw from his association with the A.·.A.·., simply notifying the Neophyte who introduced him.

There are no extravagant oaths within the A.·.A.·., for the aspirant is pledged quite simply to himself alone[24].

> 8. He shall everywhere proclaim openly his connection with the A.·.A.·. and speak of It and Its principles (even so little as he understandeth) for that mystery is the enemy of Truth. One month before the completion of his year, he shall deliver a copy of the Record of his year's work to the Neophyte introducing, and repeat to him his chosen chapter of *Liber LXV*.

There are no bonds of unnecessary secrecy within the Order. In the Aeon of Horus, all forms of esoteric knowledge should be "unveiled".

[24] The one possible exception to this is of course "the Oath of the Abyss" which constitutes one a Magister Templi (Master of the Temple).

> 9. He shall hold himself chaste, and reverent towards his body, for that the ordeal of initiation is no light one. This is of peculiar importance in the last two months of his Probation.

To quote an Arabian proverb: "He who has health has hope, he who has hope has everything".

The present writer has also observed, from the entries in his diary, that during the conclusion of his own Probationary period, chaos and mayhem seemed to be unleashed upon all aspects of his personal life. This resulted in a considerable bout of ill-health of a mental and physical nature.

> 10. Thus and not otherwise may he attain the great reward: YEA, MAY HE OBTAIN THE GREAT REWARD!

These techniques are infallible and lead to rewards of inconceivable marvel.

A MODERN INTERPRETATION OF LIBER LXI

The following historical account is based solely on Crowley's own History Lection (*Liber LXI*) and *should not* be regarded as the only valid testimony concerning the original Golden Dawn's ancestry and subsequent demise. For those readers inquisitive enough to pursue a more scholarly approach to the subject, the present writer strongly recommends the study of Dr. Francis Israel Regardie's astute work, *What You Should Know About The Golden Dawn.* Here Regardie relates his own personal experience of the G.D. system and builds on conclusions from the accounts and experiences of others.

Unfortunately so much personal animosity has entered into the debate between Golden Dawn scholars that the present writer fails to see what possible good can now come of squabbling over the origins of an occult Order whose techniques clearly work. The fact that these techniques do work - and that they can have their effectiveness verified through experimentation - should suffice as proof alone that the Golden Dawn, and hence the A∴A∴, is rooted in divine authenticity.

THE HISTORY LECTION

Many years ago (probably during 1887) a London coroner, by the name of Dr. W. Wynn Westcott, came upon a number of cipher manuscripts which purported to derive from the Rosicrucians. Westcott asked a renowned occult scholar of the time, S.L. MacGregor Mathers, to help him decipher these enigmatic documents which, when decoded, revealed a number of undeveloped rituals of a Rosicrucian nature. The manuscripts were alleged to contain the name and address of one Anna Sprengel of Nuremberg, Germany, whom Westcott claimed he wrote to and subsequently received a charter to found the Order of the Golden Dawn in Britain.

This Order operated in a semi-secret manner, and with the undeveloped rituals elaborated by Mathers, the Golden dawn began to recruit a whole host of influential (and not so influential) people.

In 1891 Anna Sprengel was rumoured to have died, and Westcott announced that her colleagues refused point blankly to encourage the British charter any further. Apparently Sprengel's associates had originally disapproved of her strategy, however, since it is the supreme rule of adepts never to interfere with the judgement of any other individual, her scheme proceeded without restraint.

According to Westcott, the German adepts claimed that the Golden Dawn had already acquired sufficient knowledge to forge its own links with the "true" overseers of the Order: "Secret Chiefs" who were purported to be superhuman beings concerned particularly with the evolution of humanity.

Some time later, Mathers asserted that he had established the appropriate magical link with these chiefs, informing the Order's members that he and two others were to govern the Golden Dawn and establish a second, more secretive, inner Order. Mathers and Westcott appointed an eminent Freemason, Dr. William Robert Woodman, to assist them in their administration, and pages of practical and theoretical occult instruction flowed from Mathers' clairvoyantly inspired pen.

These teachings, despite the curious circumstances under which they were acquired, possessed a coherent composition that was both logical and complex; however, within a very short period, Dr. Woodman died leaving Mathers and Westcott to maintain control of the Order that they had established.

For six years the Golden Dawn flourished under the management of its two founding members, until eventually - for reasons still uncertain - Westcott resigned, securing Mathers as the sole authority. The London members began to grow tired of Mathers' autocratic behaviour. Rituals were elaborated into pompous nonsense, unsuitable candidates were admitted for financial rather than intellectual reasons, and initiations became a parody of their original intent.

Inevitably scandal arose.

Then in 1900, Aleister Crowley (who was initiated into the G.D. in 1898) began to test both the capability of Mathers and the Order itself. He discovered that although Mathers was indeed a scholar and a magician of some notable distinction, his expanded ego had prevented him from attaining complete initiation. The claim of Mathers that the Secret Chiefs were solely in charge of the Order was refuted; and Crowley, by scandalous means, ensured the demise of the Order and its leader.

The fact that Crowley was far from a perfect adept himself led him to follow in the footsteps of his hero, Sir Richard Burton, where for six years he travelled to distant lands and countries, studying the world's religious doctrines and systems of initiation. This, he believed, bestowed upon him a certain grade so exalted that he was able to perceive the inadequacy of science, philosophy and religion; thereby exposing the self-contradictory nature of man's thinking faculty.

The new master returned to Britain and laid his achievements humbly at one George Cecil Jones (a former Golden Dawner of some notable ability). Jones accepted Crowley's authority and the two adepts conferred on how a new Order, free from the errors and deceits of the former, could be established.

Exalted as their degree of adeptship was, Crowley and Jones could not establish a new Order without authority. They resolved therefore to make preparations for such an event, since they knew not where to look for higher adepts than themselves. They did realise, however, that the only true way to attract the notice of such masters was to equilibrate the symbols. Therefore, under Jones's guidance, Crowley prepared all things by his esoteric science and wisdom, choosing only those symbols which were common to all systems, whilst simultaneously rejecting anything of a dogmatic nature.

To do this properly was found to be virtually impossible[25], for all language is built on history and terms like "spirit", for example, implies Scholastic Philosophy as well as the Hindu and Taoist theories regarding the breath of man.

It was therefore extremely difficult to avoid the implications of appearing biased, and for that reason they took refuge in vagueness; not in an endeavour to conceal the truth, but to warn the student against valuing the non-essential. Therefore, in the event that the Probationer comes upon the name of a God, let him not rashly assume that it refers to a specific God (other than the God that is part of himself). And if any ritual makes reference to a specific philosophy (whether Egyptian, Greek, Christian or otherwise), one must understand it as a language imperfection: a literary defect rather than a religious prejudice. Take note that many aspirants (on many levels) have fallen from the path by failing to adhere to this advice. For the essence of true wisdom has been lost in this manner in all the other systems of attainment.

Now when Crowley had finally prepared all the things under Jones' guidance, there was a period of lying fallow, since all seeds need time to germinate. They therefore continued the Great Work and within the fullness of time their seeds developed into trees of fruit, at which point they were admitted to the Eternal and invisible Order that has No Name.

-oOo-

[25] Chaos Magicians take note.

CHAPTER THREE

ESTABLISHING THE MAGICAL LINK

As previously mentioned, it is a great falsehood that one need contact a chartered A.·.A.·. Order to establish an appropriate magical link to that Order. Those who maintain otherwise are quite simply, in contemporary terms, talking bollocks. The reason for this is in fact quite straightforward. It is said that there is only one true order of initiates, to which all enlightened persons are affiliated. If this is indeed true, then it is reasonable to assume that the A.·.A.·., as founded by Crowley and Jones, is a symbolic representation of this one true Order, since its grade structure comprises of a hierarchical grade system encompassing every level of spiritual attainment. Therefore, it would also seem reasonable to conjecture that all one basically needs to do in order to become a member of this Great White Brotherhood is attain a reasonable degree of illumination by means of adhering to known magical techniques and applying a little self-discipline. As Crowley himself remarked:

> In the A.·.A.·., which is a genuinely Magical Order, there is no extravagant oaths. The candidate is pledged quite simply to himself only, and his obligation binds him merely "to obtain the scientific knowledge of the nature and powers of my own being"... When a man ceremonially affirms his connection with the A.·.A.·. he acquires the full power of the whole Order. He is enabled from that moment to do his true will to the utmost without interference.[26]

Therefore, perceiving the A.·.A.·. from this perspective - as a system offering a means of illumination to any individual courage enough to embark upon such a quest - one can discern no reason why any aspirant, of a sincere nature, cannot become affiliated to the Order by simply swearing the necessary oath and performing the appointed tasks. As Jake Stratton-Kent pointed out in his interesting document *Liber Achad*[27]:

> The A.·.A.·. is the Sanctuary Crowley sought before he found the Golden Dawn, it is the Sanctuary from which the Golden Dawn, the O.T.O.[28] and the Rosicrucians of yore emerged into the world, from which Crowley's A.·.A.·. itself emerged. If proof of membership were required, which would be absurd, then no document signed by Crowley or his successors or his imitators would be valid. Those who sense, believe or know this to be true may be members of this august body already... The real A.·.A.·., in a sense, was recreated when Regardie put the Oaths and Tasks of the Grades in the back of *Gems From The Equinox*. Those who took those Oaths and performed those Tasks are as genuine, or more so, than any charter holding claimant.

Here then follows the Oath of a Probationer:

---

[26] Crowley, A; *The Confessions Of Aleister Crowley*, page 660 (Arkana Books, 1989).

[27] This excellent piece was published within *The Equinox - British Journal Of Thelema* (Volume VII, No. 6) and is available from Kiblah Publishing, Park House, Batcombe, Dorset. The writers who produce this periodical are worthy occultists and one could far worse than subscribe to their journal.

[28] Ordo Templi Orientis or Order of the Eastern Templars. This was a more masonic magical Order which Crowley became Head of in 1925. Unfortunately its "adepts" have today fallen into the same fate as those belonging to the Hermetic Order of the Golden Dawn in Crowley's time; for like the G.D back then, the Order now utterly fails to initiate.

A.·.A.·.

The Oath of a Probationer

I, ..., being of sound mind and body, on this .... day of .... (Anno .... Sol in ....° of ....) do hereby resolve: in the presence of ...., a Neophyte of the A.·.A.·.: To prosecute the Great Work: which is, to obtain a scientific knowledge of the nature and powers of my own being.

May the A.·.A.·. crown the work, lend me of Its wisdom in the work, enable me to understand the work!

Reverence, duty, sympathy, devotion, assiduity, trust do I bring to the A.·.A.·. and in one year from this date may I be admitted to the knowledge and conversation of the A.·.A.·.!

Witness my hand ....

Motto ....

Before swearing this oath, the student is advised to digest fully all that has been previously stated. One should therefore read carefully the wording of this declaration and ensure that it is in accordance with one's own True Will. Once this has been fulfilled, the student may then proceed by following the prescribed agenda here mentioned:

1. The Preparation of the Place
   (a) Tidy the room and remove all objects which are likely to distract attention.
   (b) Set up an altar in the centre of the room (any old table will suffice) and place upon it a copy of the oath, a candle and a pen (one may also wish to include the proper apparatus for burning incense).

2. The Preparation of the Body
   (a) Ensure that one feels thoroughly clean (i.e. maybe take a shower or a bath).
   (b) Throw some salt into a separate tub of lukewarm water and cleanse oneself with the contents. Whilst doing this recite the words: "For pure will, unassuaged of purpose, delivered from the lust of result, is every way perfect."[29]
   (c) Dry oneself thoroughly and anoint the forehead, hands and feet with Oil of Abramelin[30]. Recite the words "I am uplifted in thine heart; and the kisses of the stars rain hard upon thy body."[31]

3. The Preparation of the Mind
   (a) Don the robe and proceed to the place of working.
   (b) Light the candle and incense & kneel comfortably upon the floor facing east.
   (c) Try to relax completely, removing all mundane concerns from the mind.
   (d) Remain like this for as long as feels appropriate before proceeding to the next stage.

4. The Swearing Of The Oath
   (a) Give the signs of Horus & Harpocrates[32], or perform - if one is at all familiar with the technique - the Lesser Banishing Ritual of the Pentagram.
   (b) Knock upon the alter 3-5-3 proclaiming: "Do what thou wilt shall be the whole of the Law".
   (c) Fill in the oath and read it aloud before signing it with name and motto (note: since one is taking this oath without the presence of a Neophyte, one is advised to insert the name "Aiwass" within the section pertaining to one's superior).
   (d) Knock once upon the alter affirming: "Love is the law, love under will".
   (e) Repeat the gestures as in (a) or perform the Lesser Banishing Pentagram Ritual.

-oOo-

[29] From *The Book Of The Law*, chapter I, verse 44.
[30] For Oil of Abramelin mix four parts cinnamon oil, two parts myrrh oil, one part galangal oil, and seven parts olive oil (note: rather than spend hours synthesizing the oil from scratch, one should be able to purchase this sacred fluid, ready made, from any such reputable dealer in occult paraphernalia).
[31] From *The Book Of The Law*, chapter II, verse 62.
[32] The Sign of Horus: Quickly advance the left (or right) foot approximately 12 inches, throw forward the body and let the hands (drawn back to the side of the eyes) shoot out horizontally in front of you.
The Sign of Harpocrates: Withdraw the extended foot and place the thumb of the right hand over the lips to assume a gesture of silence.

CHAPTER FOUR

THE FORMULA OF IAO

And Its Relevance Within The Period Of Probation

In the Western Esoteric Tradition there is a formula known as IAO, or Isis-Apophis-Osiris. It is in essence the formula of yoga and meditation, however its effects are far more universal, for it encompasses every magical activity that we, as humans, embark upon. As Aleister Crowley explained:

> **In beginning a meditation practice, there is always quiet pleasure**, a gentle natural growth; one takes a lively interest in the work; it seems easy; one is quite pleased to have started. This stage represents Isis. Sooner or later **it is succeeded by depression** - the Dark Night of the Soul - an infinite weariness and detestation of the work. The simplest and easiest acts become almost impossible to perform. Such impotence fills the mind with apprehension and despair. The intensity of this loathing can hardly be understood by any person who has not experienced it. This is the period of Apophis.
>
> It is followed by the arising not of Isis, but of Osiris. **The ancient condition is not restored, but a new superior condition is created**, a condition only rendered possible by the process of death.[33]

It may therefore be somewhat disturbing for the Probationer to learn that this is the path set out before him, for each of these three phases are incorporated not only into the period of Probation, but within the whole structure of the Order and each of its eleven grades. It must be remarked that from what the present writer has experienced to date, the arising of Apophis is far from a pleasurable experience. Nevertheless, it is an essential condition in the striving to attain Illumination. As Gerald Suster observed:

> Coming to know God - or the attainment of Superconsciousness or whatever sectarian terms is preferred - is for most a terrifying experience which evokes every possible dread and element of paranoia from the psyche. It takes a while to comprehend that all one's intellectual conceptions, arrived at after agonies, are still inadequate. Many find that their lives go haywire at this point, that the Universe proceeds to behave in a manner contrary to all known rules of Physics and Psychology as accepted by most learned men and women of humanity. You will probably have to face your own deepest fears - and not in your head, which will in any case be sufficiently bombarded by perplexing data, but in the events of daily living.[34]

There is, of course, very little that one can do during this relatively unpleasant phase except knuckle down and get on with the tasks; holding steadfastly all the time to the knowledge that it will eventually - upon arrival of Osiris - be over. The only encouraging words the present writer can give to those suffering these deplorable effects, is that if one persists with courage and determination, one will surely emerge like the phoenix from the fire: much stronger, remarkably tougher, and more determined and able to cope with the negative events of every day life. Heartening words, no doubt, to the Probationer whose wife has just left him one hour prior to his hernia operation.

The next six chapters, then, are arranged with respect to the aforementioned formula. With the Probationary period lasting - under ideal conditions - one whole year, one could quite simply deal with a single chapter every two months. However, as Israel

[33] *Magick*, page 167 (Guild Publishing, 1989).

[34] *The Truth About The Tarot*, page 78 (Skoob Books Publishing Ltd, 1990).

Regardie correctly pointed out in his Preface to *The One Year Manual*: "...each student represents a different problem. Each is a unique personality with his own character-structure, his own idiosyncrasies and his own way of solving problems in a certain amount of time." Therefore, unless one considers oneself to be a "standard" or "ordinary" individual (which one truly hopes is not the case), one is thus advised simply to progress by feeling alone. There are plenty of stoppage points along the way for one to gauge one's abilities. If one is performing the prescribed techniques successfully, and obtaining the desired results effectively, then one should push onward in total confidence.

With the aid of this handbook and a scrupulously well-kept Record, there should be no reason whatsoever for the Probationer to fail in his aspiration towards understanding and self-improvement.

-oOo-

CHAPTER FIVE

THE AWAKENING OF ISIS

Some Essential Instructions At The Inception Of The Work

I - The Employment Of The Magical Diary

1. It is absolutely necessary that all experiments should be recorded in detail during, or immediately after, their performance.

2. It is highly important to note the physical and mental condition of the experimenter or experimenters.

3. The time and place of all experiments must be noted; also the state of the weather, and generally all conditions which might conceivably have any result upon the experiment either as adjuvants to or causes of the result, or as inhibiting it, or as sources of error.

4. The A.'.A.'. will not take official notice of any experiments which are not thus properly recorded.

5. It is not necessary at this stage for us to declare fully the ultimate end of our researches; nor indeed would it be understood by those who have not become proficient in these elementary courses.

6. The experimenter is encouraged to use his own intelligence, and not to rely upon any other person or persons, however distinguished, even among ourselves.

7. The written record should be intelligently prepared so that others may benefit from its study.

8. The Book *John St. John* published in the first number of *The Equinox* is an example of this kind of record by a very advanced student. It is not as simply written as we could wish, but will show the method.

9. The more scientific the record is, the better. Yet the emotions should be noted, as being some of the conditions.

Let then the record be written with sincerity and care; thus with practice it will be found more and more to approximate to the ideal.

Very little need be added to Uncle Aleister's words, except that the record will become one of the most priceless books in the Probationer's collection. It will prove, time and time again, to be an invaluable source of information, and it will - at a moment's notice - help one to discover the relation between cause and effect in all of one's workings. It cannot be stressed just how important the keeping of the record is, for although this esoteric system is the aim of religion, it is equally the method of science, and must therefore be treated with the strictest scientific proficiency.

The Probationer, then, should perhaps begin the record by writing a short biography of his life so far, explaining fully his reasons for embarking upon the work, whilst at the same time capturing the essence of his aspirations. Once this has been accomplished he will be ready to commence some practical experiments.

II - The Lesser Banishing Ritual Of The Pentagram

Out of all the ceremonies a magician might perform in his lifetime, the Lesser Banishing Ritual of the Pentagram will prove undoubtedly the most useful and most used. It is performed at the onset of all magical workings and at the conclusion, and it is employed to disperse negativity in whichever form it manifests. For this reason it will have especial relevance during the period of Apophis. Therefore, rather than wait until the onset of this phase is upon the Probationer, the present writer strongly urges him to acquire familiarity with the technique immediately. Obsession is one of the most perilous fiends that one will have to magically contend with, hence pains must be taken to avoid it at all costs.

The first task the magician has to accomplish is mastery over his immediate environment[35]; therefore the Pentagram - being a symbol of the phenomenal universe - allows one to achieve just that. A successful performance of the rite will result in an undoubted feeling of clearness, which assists one to effectively establish a sound magical environment suitable for all kinds of ceremonial working.

Here then follows two versions of the Lesser Banishing Ritual of the Pentagram, of which the first is the more traditional. The second version is my own, more Thelemic, reproduction which has been designed solely because nobody else seems to have bothered to do so.

The Traditional Lesser Banishing Ritual Of The Pentagram

Face East.

1. Touch the forehead and say "Ateh" (Unto Thee).
2. Touch the breast and say "Malkuth" (The Kingdom).
3. Touch the right shoulder and say "ve-Geburah" (and the Power).
4. Touch the left shoulder and say "ve-Gedulah" (and the Glory).
5. Clasp the hands upon the breast and say "le-Olahm, Amen" (To the Ages, Amen).
6. Trace the Banishing Pentagram of Earth (see figure 18 on page 39) and vibrate "IHVH" (pronounced: Yeh-Hoh-Vau).
7. Turn to the South and repeat, but vibrate "ADNI" (pronounced: Ah-Don-Ay).
8. Turn to the West and repeat, but vibrate "AHIH" (pronounced: Eh-Huh-Yeh).
9. Turn to the North and repeat, but vibrate "AGLA" (pronounced: Ah-Geh-Lah).
10. Extend the arms in the form of a cross and say:
11. "Before me, Raphael!"
12. "Behind me, Gabriel!"
13. "On my right hand, Michael!" (pronounced: Mi-Kay-El).
14. "On my left hand, Auriel!"
15. "For about me flames the Pentagram!"
16. "And in the column stands the six-rayed Star."

17 - 21. Repeat 1 - 5, the Qabalistic Cross.

The Thelemic Lesser Banishing Ritual Of The Pentagram

1. Facing East (towards Boleskine[36]); stand upright, feet together with hands clasped over the breast.

---

[35] The Neophyte will find this matter substantially significant.

[36] Boleskine House is situated on the south-eastern shore of Loch Ness, 17 miles from Inverness at Latitude 57.14N and Logitude 4.28W. In Thelemic Magick one should regard Boleskine as the Kiblah, that is to say: the point to which one turns to make prayer (i.e. Mecca is the Kiblah of the Mohammedan).

2. Breathing in deeply, imagine a surge of energy beginning directly above the head, proceeding slowly down the body, into the lungs, down past the solar plexus & genitals, and on towards the feet.
3. The moment that it appears to reach the feet, quickly advance the left foot about 12 inches whilst throwing forth the body so that you are standing typically in the position of the god Horus. Bellow the word: "Bahlashti! Ompehda!"
4. Withdraw the left foot, placing the thumb of the right hand over the lips, assuming the form of the god Harpocrates.
5. Trace, with the thumb of the right hand, a circle about the head and say "NUIT".
6. Touch the heart and say "AIWAZ".
7. Touch the genitals and say "HADIT".
8. Touch the right shoulder and say "THELEMA".
9. Touch the left shoulder and say "AGAPE".
10. Clasp the hands over the breast and vibrate "AUMGN".
11. Imagine yourself as an Ankh of Light, where your head forms the loop and your body forms the Tau.
12. Placing the thumb of the right hand between its index and medius, trace the Banishing Pentagram of Earth whilst vibrating the formula "THE-RI-ON".
13. Turning towards the North, repeat as in 12 but vibrate "HOOR-PA-KRAAT".
14. Turning towards the West, repeat but vibrate "BA-BA-LON".
15. Turning towards the South, repeat as before but vibrate "RA-HOOR-KHU".
16. Turning once more to face Boleskine, extend the arms and legs in the form of a pentagram, and say:
17. "Before me, Iunges!"
18. "Behind me, Teletarche!"
19. "On my right hand, "Synoches!"
20. "On my left hand, "Daimonos!"
21. Visualise these entities as pillars of brilliant white light extending infinitely in height and depth, then say:
22. "For about me flames the Pentagram!"
23. "Whilst above and below me shines the six-rayed Star!"
24. Imagine the Unicusal Hexagram shining exactly as described (see page 39).

25 - 31. Repeat steps 5 - 11, the Qabalistic Ankh.

Crowley, in 1899, purchased Boleskine House as a suitable place to perform the Sacred Magic of Abra-Melin the Mage (a magical operation employed for the attainment of the Knowledge and Conversation of one's Holy Guardian Angel). This operation was only partially performed, nevertheless, it did prepare him sufficiently enough in order to form a solid foundation for the reception of *The Book Of The Law*, and for this reason Boleskine was considered sacred by Crowley thereafter. Oddly enough, *The Book Of The Law* makes a number of references to a "holy place" or "Kiblah"; and in particular "...the name of thy house 418". Later, Crowley discovered that the word "Boleskine", by Hebrew Gematria, adds up to 418, which is also the number of "Aiwass" and "Abrahadabra".

When performing a Thelemic ceremony, one should assume Boleskine as one's East. Thus North, South and West will alter in accordance with this. The reasons for making these alterations are quite diverse. Some magicians, for example, consider this to be of great religious significance, whilst others regard it as a method of altering our conscious perception of the universe around us. The present writer agrees with both these reasons, inasmuch that in the Aeon of Horus we must regard ourselves as Solar Beings and strive to avoid falling back into the consciousness of the previous Aeon, where our ancestors based their religious ideas on the rising and setting of the sun (the slain and risen god of whom Osiris, Adonis, Attis, Dionysus and Jesus Christ are all based).

It will be noted that this ritual incorporates a number of magical techniques of which some Probationers might never before have come across (notably: the assumption of God-forms, the vibration of words, and creative visualisation).

Here then follows a number of diagrams that the layperson should become familiar with. One should also familiarise oneself with the some of the divine names given, and what they represent. Once this has been achieved, one will be ready to follow the exercises prescribed thereafter.

Figure 1
The Sign Of A Neophyte

The God Set Fighting
Earth (1°=10□)

Figure 2
The Sign Of A Zelator

The God Shu
Air (2°=9□)

Figure 3
The Sign Of A Practicus

The Goddess Auramoth
Water (3°=8□)

Figure 4
The Sign Of A Philosophus

The Goddess Thoum-aesh-neith
Fire (4°=7□)

Figure 5
The Sign Of An Adeptus Minor

The Rending Of The Veil
Active Spirit ($5^{\circ}=6^{\square}$)

Figure 6
The Sign Of An Adeptus Major

The Closing Of The Veil
Passive Spirit ($6^{\circ}=5^{\square}$)

Figure 7
The Sign Of An Adeptus Exemptus

Osiris Slain
+ - The Cross Water ($7^{\circ}=4^{\square}$)

Figure 8
The Sign Of A Magister Templi

Isis Mourning
L - The Swastika Water ($8^{\circ}=3^{\square}$)

Figure 9
The Sign Of A Magus

The God Typhon
V - The Trident Water (9°=2▫)

Figure 10
The Sign Of An Ipsissimus

Osiris Risen
X - The Pentagram Water (10°=1▫)

Figure 11
The Sign Of The Enterer

The God Horus

Figure 12
The Sign Of Silence

The God Harpocrates

Figure 13
The Sign Of Puella

Figure 14
The Sign Of Puer

N

Figure 15
The Sign Of Mulier

O

Figure 16
The Sign Of Vir

X

Figure 17
The Sign Of Mater Triumphans

Figure 18
The Banishing Pentagram Of Earth

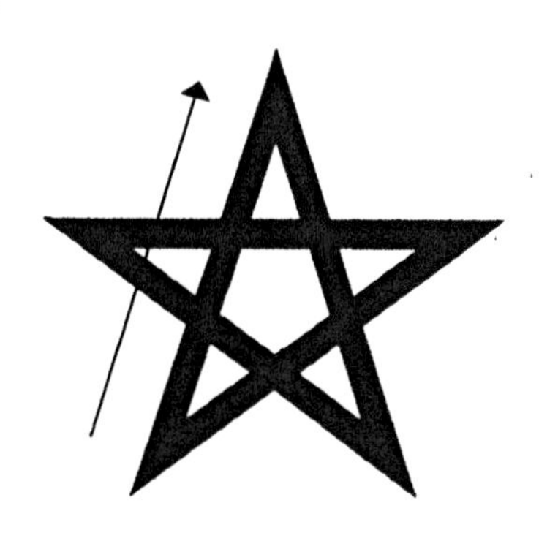

Figure 19
The Hexagram

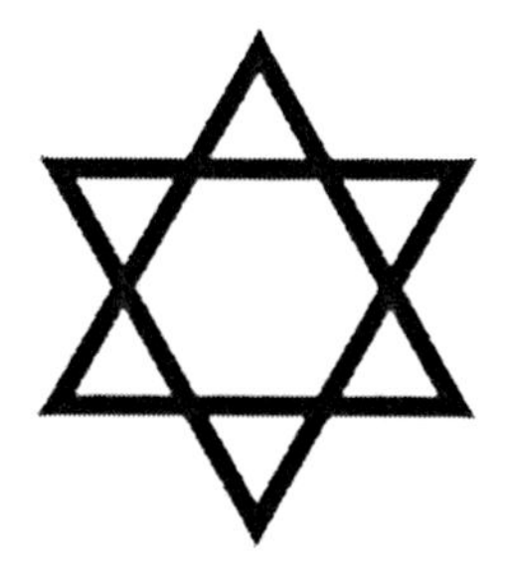

Figure 20
The Unicursal Hexagram

Formula	Description
IHVH	Pronounced Yeh-Ho-Vau, this is the formula of Tetragrammaton; Yod (I), Heh (H), Vau (V), Heh (H), the Ineffable Name (Jehovah) of the Hebrews. Represents the four elements: Fire, Water, Air, Earth respectively.
ADNI	Pronounced Ah-Don-Ay, this formula is best described in the Gnostic sense by Crowley in *Liber Samekh*: "My Lord! My secret self beyond self, Hadith, All Father! Hail, ON, thou Sun, thou Life of Man, thou Fivefold Sword of Flame! The Goat exalted upon Earth in Lust, thou Snake extended upon Earth in life! Spirit most holy! Seed most wise! Innocent Babe. Inviolate Maid! Begetter of Being! Soul of Souls! Word of all Words, Come forth, most hidden Light!"
AHIH	Pronounced Eh-Huh-Yeh, this name simply means "existence" and is a title attributed to Kether.
AGLA	Pronounced Ah-Geh-Lah, this word is a Notariqon formed from the initials of the sentence "Ateh Gibor Leolahm Adonia" meaning "To Thee be the Power unto the Ages, o my Lord."

Auriel	The Archangel of Earth.
Raphael	The Archangel of Air.
Gabriel	The Archangel of Water.
Michael	Pronounced Mi-Kay-El; the Archangel of Fire.
NUIT	Traditionally the Egyptian sky goddess, she represents the infinite circumference of the expanded universe (see *The Book Of The Law*, chapter I).
HADIT	Depicted as a winged globe of light, this god symbolises the microcosmic point within the centre of the universe (see *The Book Of The Law*, chapter II).
RA-HOOR-KHUIT	The child of Nuit and Hadit, he is quite simply the manifested universe and the range in which it operates (i.e. between the infinitely small and the infinitely large). Since this results in a Unity which includes and heads all things, Ra-Hoor-Khuit (or more correctly Heru-Ra-Ha) symbolises both the active and passive nature of existence. The name Ra-Hoor-Khuit is therefore associated with the concept of extroversion, whereas Hoor-Paar-Kraat (the passive side of Heru-Ra-Ha) is a title suggesting introversion.
AIWAZ	Also spelt AIWASS, this is the name given to the form of intelligence who dictated *The Book Of The Law* to Aleister Crowley in 1904. Describing himself as the "minister of Hoor-paar-kraat", Aiwaz was identified by Crowley as his own Holy Guardian Angel.
THELEMA	The Greek word for "Will", this word has a Gemateric value of 93 and was employed by Crowley to describe the formula of "Do what thou wilt".
AGAPE	The Greek word for "Love", this too possesses a Gematric value of 93. One is therefore instantly reminded of the commandment: "Love is the law, love under will" (from *The Book Of The Law*, chapter I, verse 57).
AUMGN	Variation of the Hindu mantra AUM it is described as the root sound of Creation. Crowley added the compound letter GN to the formula since he felt that by so doing it symbolised more accurately the course of spiritual existence. As he wrote in *Liber Samekh*: "A is the formless Hero; U is the six-fold solar sound of physical life, the triangle of Soul being entwined with that of the Body; M is the silence of 'death'; GN is the nasal sound of generation & knowledge." Its Gematric value is 93 also.
THERION	The Greek title for "Beast", this word has a Hebrew Gematric value of 666. Readers open minded enough to pursue and intelligent interpretation of this number, without religious bias, will realise that it represents the highest concept of man's evolution, in the sense that in order to become truly free as individuals we must realise our divine nature as well as our animal instincts.
BABALON	The ultimate concept of womanhood, keeper of the Holy Graal.
IAO	As previously explained in Chapter Four of this manual: I is Isis, Nature ruined by A, Apophis the Destroyer, and then restored by O, Osiris the Redeemer. It is therefore the same idea that one finds

	expressed in the Rosicrucian formula of the Trinity. Occasionally within a Thelemic setting one will find this formula prefixed and suffixed by a V or F (the Hebrew letter Vau), thus giving us the formula VIAOV. Crowley did this for a number of reasons; firstly to give it a more solar-phallic nature (since Vau corresponds to the Solar 6); and secondly to give the whole formula a Gematric value of 93 (thus equating it with the current of the Aeon).
ARARITA	A name of God derived from the Notariqon of the Hebrew sentence "One is His Beginning; One is His Individuality; His Permutation One."
ABRAHADABRA	The Word of the Aeon of Horus. With its five identical letters and six diverse it represents the conjunction of the microcosm (the Pentagram) with the macrocosm (the Hexagram), thus symbolising the accomplishment of the Great Work.

III - Fundamental Training In Practical Magick

1. Practice the Qabalistic Cross and imagine it manifesting itself within you as streams of brilliant white light.
2. Employ this technique at the onset and conclusion of all magical experiments.
3. Practice tracing the Banishing Pentagram of Earth and visualise the symbol blazing brilliantly in front of you. Try to feel the effects of how it might expel any negativity you may be experiencing.
4. Practice the vibration of names and incorporate them into the tracing of the pentagrams. One should try to get the maximum sonic resonance from the word selected, keeping the voice as sonorous as possible (try "humming" out the name as this will give you a better understanding of what is meant by the term "vibrate").
5. Practice the assumption of magical God-forms, and try to feel the effects that these produce. Do they characterise the forces that they represent? Allow the whole concept of the God to indwell in you and all that it represents.
6. Practice various sequences of magical gestures (i.e. the signs of L.V.X., the signs of N.O.X., and the signs of Horus and Harpocrates) and note the results that manifest.
7. Practice the whole Lesser Banishing Ritual of the Pentagram and incorporate it into your daily routine. Perform it at least twice daily and note any effects.[37]
8. Perform the ritual for a specific purpose and observe how it removes any opposing forces that surround that purpose.
9. Make a careful note of all these experiments and include the results in the magical diary.

[37] In fact, at the Probationary level, one should endeavour to perform the ritual as often as possible, since it is the quickest and most effective may to balance one's thoughts, emotions and imagination. However, as Crowley correctly observed in *Magick*: "...as the student advances to Adeptship the banishing will diminish in importance, for it will no longer be so necessary. The Circle of the Magician will have been perfected by his habit of Magical work. In the truest sense of the word, he will never step outside the Circle during his whole life. But the consecration, being the application of a positive force, can always be raised to a closer approximation to perfection. Complete success in banishing is soon attained; but there can be no completeness in the advancement to holiness."

IV - Fundamental Training In The Preliminaries Of Yoga

In simple terms "Yoga" means "Union"; therefore rather than bore any sane person to the point of madness, let us skip over all the intellectual groundwork and get on with the practices. This is, after all, a practical manual designed solely to take the reader through the period of Probation. If one is at all serious about the work, one should have previously studied the relevant material in either Crowley's *Eight Lectures On Yoga* or *Magick (Book Four)*, part 1. In which case the reader will hopefully concur that the matter is relatively straightforward and not at all as complex as demented, new age, middle-class housewives would have one believe[38].

For the present then, let us concentrate on Posture (Asana), Regularisation of Breath (Pranayama) and some Basic Meditation techniques (Pratyahara & Dharana). Although it is not necessary for one to gain mastery in any of these practices, one would certainly benefit considerably by devoting much of the Probationary period to acquiring a certain amount of adeptness in the following exercises:

1. Remove all garments that might interfere with the experiment and try to sit perfectly still with the muscles taut for long periods.
2. With eyes closed, kneel in the Dragon position[39] (buttocks resting on the heels, toes turned back, spine and head straight, hands on thighs) and take note of anything peculiar that appears to take place.
3. Time yourself properly for each of practice and record any relevant information such as itching, cramps, movement or interruptions[40]. Try to become fully aware of the body.
4. Once a certain amount of proficiency has been achieved in the practice, introduce a steady rhythm to your breathing; inhaling for four seconds and exhaling for the same (one might also try holding one's breath for four seconds at the conclusion of each breath sequence).
5. Attempt various longer breath cycles and find out with which you feel most comfortable[41]. Note any peculiarities.
6. Move onto some meditation practices, first of all by observing the contents of the mind and then tracing back the origins of each thought that presents itself.
7. In order to avoid the possibility of obsession, decide beforehand exactly how long you wish the duration of your introspection to last for, and adhere to that decision.
8. Try to become aware of all the "stuff" that is floating about your psyche. Do not be alarmed if some of the thoughts prove disturbing - this is a perfectly commonplace reaction.

---

[38] As Crowley summed up the situation in eight words: "Sit still. Stop thinking. Shut up. Get out."

[39] In *Magick (Book Four)* Crowley gives four such meditation postures for the student to choose from: The God, The Dragon, The Ibis and The Thunderbolt. In the present writer's opinion, although The God will prove to be even easier than The Dragon position, it does require the use of a chair, and this might not always be readily available at hand (even if these contraptions are immensely popular). As for The Ibis and The Thunderbolt, one often wonders just how far Crowley used to take his practical jokes.

[40] Although one such interruption that might occur regularly is an obsession with checking the time.

[41] In order to fully attain the grade of Zelator, Crowley recommended that one should be able to maintain (for a full hour) a breath cycle comprising of a 30 second exhalation and 15 second inhalation. Despite what the present writer believed at the beginning of his training in occultism, such a breath cycle is indeed quite possible with practice. In fact one may wish to try Pranayama after indulging in a little cannabis.

9. Once a certain amount of calm is established within the mind, try condensing your thoughts to a single object (i.e. should you chose to meditate on a rose, then simply conjure up such an image in your head and endeavour to experience all that it represents).
10. Observe the imagined object. Listen to the imagined object. Smell, touch and taste the imagined object. In simple terms, unite yourself fully with the object and avoid the intrusion of renegade thoughts.
11. Make a careful not of all these experiments and include the results in the magical diary.

V - Some Additional Practices To Induce Magical Realisation

The following rituals were composed by Crowley with the intention of inducing within the student a greater understanding of his individual nature. Like the Lesser Banishing Pentagram Ritual, each should be performed daily and incorporated into the structure of everyday life.

Will - A Magical Dialogue To Be Performed Before A Meal

Take a knife and knock 3-5-3.
A: "Do what thou wilt shall be the whole of the Law."
B: "What is thy Will?"
A: "It is my Will to eat and drink."
B: "To what end?"
A: "That my body may be fortified thereby."
B: "To what end?"
A: "That I may accomplish the Great Work."
B: "Love is the law, love under will."
A: "Fall to!"
Knock once.
(Note: the above may be adapted as a monologue).

Liber Resh - The Fourfold Solar Adorations

1. At Dawn, greet the Sun in the East, and make the following adoration giving the sign of your grade[42]:

HAIL UNTO THEE WHO ART RA IN THY RISING,
EVEN UNTO THEE WHO ART RA IN THY STRENGTH,
WHO TRAVELLEST OVER THE HEAVENS IN THY BARK,
AT THE UPRISING OF THE SUN.
TAHUTI STANDETH IN HIS SPLENDOUR AT THE PROW,
AND RA-HOOR ABIDETH AT THE HELM.
HAIL UNTO THEE FROM THE ABODES OF NIGHT!

[42] The sign of a Neophyte (Set Fighting) will suffice, as there happens to be no sign officially designated to the Probationer.

2. Also at Noon, greet the Sun towards the South, and make the following adoration giving the sign of your grade:

HAIL UNTO THEE WHO ART AHATHOOR IN THY TRIUMPHING,
EVEN UNTO THEE WHO ART AHATHOOR IN THY BEAUTY,
WHO TRAVELLEST OVER THE HEAVENS IN THY BARK,
AT THE MID-COURSE OF THE SUN.
TAHUTI STANDETH IN HIS SPLENDOUR AT THE PROW,
AND RA-HOOR ABIDETH AT THE HELM.
HAIL UNTO THEE FROM THE ABODES OF MORNING!

3. Then at Dusk, greet the Sun in the West, and make the following adoration giving the sign of your grade:

HAIL UNTO THEE WHO ART TUM IN THY SETTING,
EVEN UNTO THEE WHO ART TUM IN THY JOY,
WHO TRAVELLEST OVER THE HEAVENS IN THY BARK,
AT THE DOWN-GOING OF THE SUN.
TAHUTI STANDETH IN HIS SPLENDOUR AT THE PROW,
AND RA-HOOR ABIDETH AT THE HELM.
HAIL UNTO THEE FROM THE ABODES OF DAY!

4. Finally at Midnight, greet the Sun by facing North, and make the following adoration giving the sign of your grade:

HAIL UNTO THEE WHO ART KHEPHRA IN THY HIDING,
EVEN UNTO THEE WHO ART KHEPHRA IN THY SILENCE,
WHO TRAVELLEST OVER THE HEAVENS IN THY BARK,
AT THE MIDNIGHT HOUR OF THE SUN.
TAHUTI STANDETH IN HIS SPLENDOUR AT THE PROW,
AND RA-HOOR ABIDETH AT THE HELM.
HAIL UNTO THEE FROM THE ABODES OF EVENING!

5. At the conclusion of each of these brief invocations, one should give the sign of Harpocrates and make the following affirmation[43]:

UNITY UTTERMOST SHOWED!
I ADORE THE MIGHT OF THY BREATH,
SUPREME AND TERRIBLE GOD,
WHO MAKEST THE GODS AND DEATH
TO TREMBLE BEFORE THEE:-
I, I ADORE THEE!

APPEAR ON THE THRONE OF RA!
OPEN THE WAYS OF KHU!
LIGHTEN THE WAYS OF KA!
THE WAYS OF THE KHABS RUN THROUGH

[43] One may also wish to meditate for a short period.

TO STIR ME OR STILL ME!
AUM! LET IT FILL ME!

THE LIGHT IS MINE; ITS RAYS CONSUME
ME: I HAVE MADE A SECRET DOOR
INTO THE HOUSE OF RA AND TUM,
OF KHEPHRA AND OF AHATHOOR.
I AM THY THEBAN, O MENTU,
THE PROPHET ANKH-AF-NA-KHONSU!

BY BES-NA-MAUT MY BREAT I BEAT;
BY WISE TA-NECH I WEAVE MY SPELL
SHOW ME THY STAR-SPLENDOUR, O NUIT!
BID ME WITHIN THINE HOUSE TO DWELL,
O WINGED SNAKE OF LIGHT, HADIT!
ABIDE WITH ME, RA-HOOR-KHUIT!

The Mantra Gayatri – The Lunar Adoration

The following mantra should be performed daily at the appearance of the Moon[44] and must be continually vibrated until one's attention ceases to waver.

AUM! TAT SAVITUR VARENYAM
BHARGO DEVASYA DIMAHI
DHIYO YO NA PRATYODAYAT.[45]

All of these practices should be adhered to rigorously throughout the period of Probation (if not throughout the whole of one's life); for although, at first, their performance might seem ever so slightly silly, it will become increasingly evident that there is a serious and sound reason behind all the apparent mumbo-jumbo, which is: **to establish a conscious link between the subject and the universe.**

After, at least, two months of familiarising oneself with these techniques, one may then proceed to the next chapter of this manual. By which time one should have committed all of the practical material to memory and grasped a fair understanding of what each practice represents.

-oOo-

[44] That is to say, at the Moon's rising. Such information can be acquired either from a daily newspaper or, more conveniently, from an astronomical ephemeris.

[45] *Magick*, page 17 (Guild Publishing, 1989). The mantra in question translates to read "O! Let us strictly meditate on the adorable light of that divine Savitri (the interior Sun, etc). May she enlighten our minds!"

CHAPTER SIX

SUSTAINING THE MOMENTUM
Some Further Instruction For The Period Of Isis

Unlike the practices in the previous chapter, the following exercises need not be performed indefinitely. Nevertheless, this does not mean that one can simply ignore them, since each will prove invaluable under the influence of Apophis.

I - The Lesser Banishing Ritual Of The Hexagram

This ceremony should be performed at the conclusion of the Lesser Banishing Ritual Of The Pentagram, and is employed to intensify the banishing effect. Unlike the microcosmic formula of the Pentagram, the Hexagram symbolises the macrocosm and will, for this reason, make a greater impression upon the magician's conscious mind.

One is reminded of how at the completion of the Pentagram Ritual once visualises the Hexagram above and below, whilst quietly affirming one's power over the material universe. In the Hexagram Ritual, one goes a stage further and affirms superiority over the forces that appear to be concealed from any materialistic comprehension.

As in the previous chapter, there here follows two versions of the ritual in question. If one is relatively comfortable performing the more traditional Pentagram Ritual then one may proceed forth with this ceremony in the more traditional manner[46].

Schematics for tracing the Hexagrams are shown at the end of this chapter.

The Traditional Lesser Banishing Ritual Of The Hexagram

1. Facing East, stand upright, feet together, left arm at side, right arm across the body holding the proper weapon upright in the median line.
2. Say:
   "I.N.R.I."

---

[46] It will be noted that in the Traditional Hexagram Ritual the elements corresponding to the compass points appear to differ from that of the Pentagram Ritual. This is not a mistake. As previously pointed out, in the Hexagram Ritual one tries to raise one's consciousness to that of the macrocosm which, in turn, alters one's perspective of how the universe is perceived. In the Pentagram Ritual, the magician stands as a man surrounded by the material world; perceiving Air in the East, Fire in the South, Water in the West, and Earth in the North. Whilst in the Hexagram Ritual, the magician stands as the Sun surrounded by the belt of the zodiac; perceiving the Cardinal signs: Aires-Fire in the East, Capricorn-Earth in the South, Libra-Air in the West, and Cancer-Water in the North.

In the Thelemic Rituals then, one actually does away with all these Old Aeonic terms by assuming Boleskine as the East and circumbulating in an anti-clockwise direction. The points then become as follows: Taurus-Earth in the East, Aquarius-Air in the North, Scorpio-Water in the West, and Leo-Fire in the South.

Is Crowley taking the piss? No, for all he has actually done is slightly alter our perception of reality, by turning us upside down and utilising the Fixed signs of the zodiac as opposed to the Cardinal signs. Since, from the Sun's point of view, there is no up or down, right or wrong, Crowley has therefore striven to improve our solar-perception by taking matters one stage further than is pursued in the Traditional Hexagram Ritual. By doing this, the Sun has now become the microcosm, whilst the Universe is perceived as the macrocosm. This then is the reason why I have employed the signs of N.O.X. in the Thelemic Hexagram Ritual as opposed to the signs of L.V.X. in the more Traditional version; for N.O.X. (night) is the Universe, whereas L.V.X. (light) is the Sun.

"Yod. Nun. Resh. Yod."
"Virgo, Isis, Mighty Mother."
"Scorpio, Apophis, Destroyer."
"Sol, Osiris, Slain and Risen."
"Isis, Apophis, Osiris, IAO."

3. Extend the arms in the form of a cross and say "The Sign of Osiris Slain."
4. Raise the right arm to point upwards, keeping the elbow square, and lower the left arm to point downwards, keeping the elbow square, whilst turning the head over the left shoulder looking down so that the eyes follow the left forearm, and say "The Sign of the Mourning Of Isis."
5. Raise the arms at an angle of sixty degrees to each other above the head, which should be thrown back, and say "The Sign of Apophis & Typhon."
6. Bow the head, cross the arms on the breast, and say "The Sign of Osiris Risen."
7. Extend the arms as in 3 and cross again as in 6, saying "L.V.X., Lux, The Light of the Cross."
8. Trace the Banishing Hexagram of Fire in the East and vibrate "ARARITA."
9. Trace the Banishing Hexagram of Earth in the South and vibrate "ARARITA."
10. Trace the Banishing Hexagram of Air in the West and vibrate "ARARITA."
11. Trace the Banishing Hexagram of Water in the North and vibrate "ARARITA."
12. Repeat steps 1 - 7, the formula of L.V.X.

The Thelemic Lesser Banishing Ritual Of The Hexagram

1. Facing East (towards Boleskine); stand upright, feet together with hands clasped over the breast.
2. Intone the formula "VIAOV" then say:
   "One is The Beginning!"
   "One is Thy Spirit!"
   "Thy Permutation is One!"
3. Cry out:
   "NUIT!"
   "HADIT!" and giving the sign of Horus:
   "RA-HOOR-KHUIT!"
4. Withdraw the left foot, placing the thumb of the right hand over the lips to assume the God Harpocrates.
5. Give the signs of N.O.X. as follows:
   Puella
   Stand with feet together, head bowed, left hand shielding the Muladhara Cakkra and right hand shielding the breast.
   Puer
   Stand with feet together and head erect. Raise the right hand (the thumb extended at right angels to the fingers) so that the forearm is vertical and at right angles to the upper arm. Drop the left hand so that it rests at the junction of the thighs (the thumb extended forward with the fingers clenched).

Mulier
Separate the feet widely, throwing back the head and raising the arms so as to suggest a crescent.
Vir
Stand, feet together with head bowed and pushed out. With the fingers clenched and the thumbs thrust forward, hold the hands to the temples as if to symbolise the butting of a horned beast.
Mater Triumphans
Stand with feet together and left hand curved as if to support a child. Take the thumb and index finger of the right hand and pinch the nipple of the left breast, saying "N.O.X., Nox, the Manifestation of Night."

6. Placing the thumb of the right hand between its index and medius, trace the Banishing Unicursal Hexagram of Sol whilst vibrating the formula "ARARITA".
7. Turn to the North and repeat as in 6.
8. Turn to the West and repeat as in 6.
9. Turn to the South and repeat as in 6.
10. Repeat the steps 1 - 5 and close with the Sign of Silence.

II - Strengthening The Will

It is of the utmost importance for the Probationer to improve the strength of his Will; for as Crowley correctly observed: "There is no power which cannot be pressed into the service of the Magical Will: it is only the temptation to value that power itself which offends."[47] This then, I have found to be of the utmost truth, for it is in fact our own limited perception of our faculties which actually confines and restricts our omnipotent capabilities.

From the moment we are born, society (perhaps unintentionally) indoctrinates us into a specific way of thinking; a way which is safe and easy for us to accept, a way which we all assume is the only and correct one, since up until now we appear to have managed quite sufficiently. The fact remains, however, that if we could actually perceive everything that presently exists around us (from radio waves to ultraviolet light and beyond) each and everyone of us would go stark raving mad without proper training. It should be fully understood then, that there are far more things in the Universe than we will possibly ever comprehend. However, it is human nature to strive towards comprehension, and this, in itself, is a striving towards Godhead.

How then does one go about discerning one's full potential? Well, first of all we must re-educate ourselves by the practice of various exercises. In *Liber III vel Jugorum*[48] Crowley gives excellent examples of the type of practices we are currently looking for, where he suggests we should avoid doing a specific custom for a prearranged period of time (whether this be crossing the legs or saying the word "the"), and to painfully remind ourselves whenever we fail to adhere to the conditions we have set.

Crowley's method of education is in fact rather sever, since he instructs the student to cut himself sharply upon the forearm with a razor so that he might maintain within himself a state of near constant vigilance. To the present writer, this sounds like

[47] *Magick*, part II, page 62 (Guild Publishing, 1989).
[48] *Magick*, Appendix VII, page 491 (Guild Publishing, 1989).

perfect idiocy, and in these plague-ridden times the practice is not at all recommended. In *The One Year Manual*, Israel Regardie suggests a much more sensible approach whereby one administers a light electric shock by means of some Joke Shop gadget. This is perhaps playing things a little too safe, however it is certainly along the right lines. One has found that by applying a fully charged 9 volt battery to one's tongue, one can administer a relatively safe and most effective form of discipline. It must be remarked, however, that one is not trying to torture oneself in any way, but merely to shock one's system into a state of continual awareness. By methods such as these one's will-power will undoubtedly increase dramatically, allowing one to grasp a greater comprehension of one's full potential.

Here then are some exercises for the Probationer to perform. One should set a specific time limit for each experiment and see that it is properly adhered to (the present writer has observed that anything under a week tends to prove fairly ineffective).

1. Avoid saying a specific word that is common to everyday speech.
2. Avoid making a specific gesture that is common to everyday use.
3. Avoid discussing a certain subject that you know will inevitably arise.
4. Avoid eating a certain food that is customary to your dietary routine.
5. Avoid thinking a certain thought that is typical to the way you reflect.
6. Perform various combinations of these exercises until you can function under all at once.
7. Each time a circumstance arises where the rule is undoubtedly broken, administer 9 volts to the tongue for a minimum for 5 seconds.
8. Further: set yourself some challenging tasks and ensure that each is accomplished.

III - Fundamental Training In The Techniques Of Clairvoyance

The Probationer should strive to increase his powers of physical clairvoyance. This must be done in the most scientific manner.

By continually repeating the following experiments, one will be able to judge that there is merit in this peculiar branch of study.

Although at first the results might be poor, given time one should find that one's success rate increases significantly. Be extremely vigilant not to cheat yourself in any way.

1. Take a Tarot pack and select a card without looking.
2. Try to name the suit of that card using only your imagination[49].
3. Once you have become relatively proficient in the practice, repeat the experiment but try to discern the card's nature (i.e. Which part of the Tree of Life does it pertain to?).
4. Eventually proceed further, endeavouring to name the actual card.
5. Tabulate the results of these experiments and record them in the diary.

IV - Invoking & Banishing Various Forces Of Nature

[49] This may be relatively difficult where the Trumps are concerned, since the nature of each card in this suit is quite unique to itself and rarely has any relation to any of the other cards in the Major Arcana. Give yourself a point for discerning the right kind of energy (i.e. Fire/Wands equates rather well with Mars/The Tower).

At the end of this chapter (pages 51 - 53), one will find various diagrams showing the proper means by which one can invoke and banish the forces of nature. These should be committed fully to memory in order for them to have an increased and worthwhile effect. Also, the aspirant will find these immensely valuable during the phase of Apophis, for should he find himself confronted by a specific form of negativity, he will have the means by which to neutralise it.

The following experiments should be practiced regularly until a certain understanding and proficiency has been achieved. All of these practices should begin and end with the Banishing Rituals in order to avoid the occurrence of obsession.

1. Invoke specific elemental forces using the Lesser Ritual Of The Pentagram (i.e. instead of tracing the banishing pentagram of Earth at each quarter, replace it with the element you wish to invoke; remembering, of course, to initiate the invocation with the appropriate Pentagram of Spirit[50]). Terminate each Pentagram by giving the proper salutation (e.g. when invoking Fire, assume the sign of Thoum-aesh-neith as shown earlier).
2. Invoke a specific planetary force using the Lesser Ritual Of The Hexagram. Employ "ARARITA" for the Hexagram, vibrating the God name for the sigil. Should it be your personal preference, use the Unicursal variant.
3. Invoke a specific zodiacal force by the method described in 2. Use the appropriate ruling planet[51], before the sigil of the sign.
4. Also experiment with the banishing of these forces and observe any peculiarities.
5. Make a careful note of all these practices within the magical diary.

Once a certain amount of familiarity has been established with these techniques, one may then proceed to the next chapter.

-oOo-

[50] Active Spirit for Fire & Air; Passive Spirit for Water & Earth.

[51] The planets and the signs that they rule are as follows: Aries - Mars; Taurus - Venus; Gemini - Mercury; Cancer - Luna; Leo - Sol; Virgo - Mercury; Libra - Venus; Scorpio - Pluto (originally Mars); Sagittarius - Jupiter; Capricorn - Saturn; Aquarius - Uranus (originally Saturn); Pisces - Neptune (originally Jupiter).

PENTAGRAMS

PASSIVE		ACTIVE	
INVOKING	**BANISHING**	**INVOKING**	**BANISHING**
SPIRIT (PASSIVE)		**SPIRIT (ACTIVE)**	

Formula: *AGLA*		Formula: *AHIH*	
WATER		**FIRE**	

Formula: *AL*		Formula: *ALHIM*	
EARTH		**AIR**	

Formula: *ADNI*		Formula: *IHVH*	

HEXAGRAMS (ELEMENTAL)

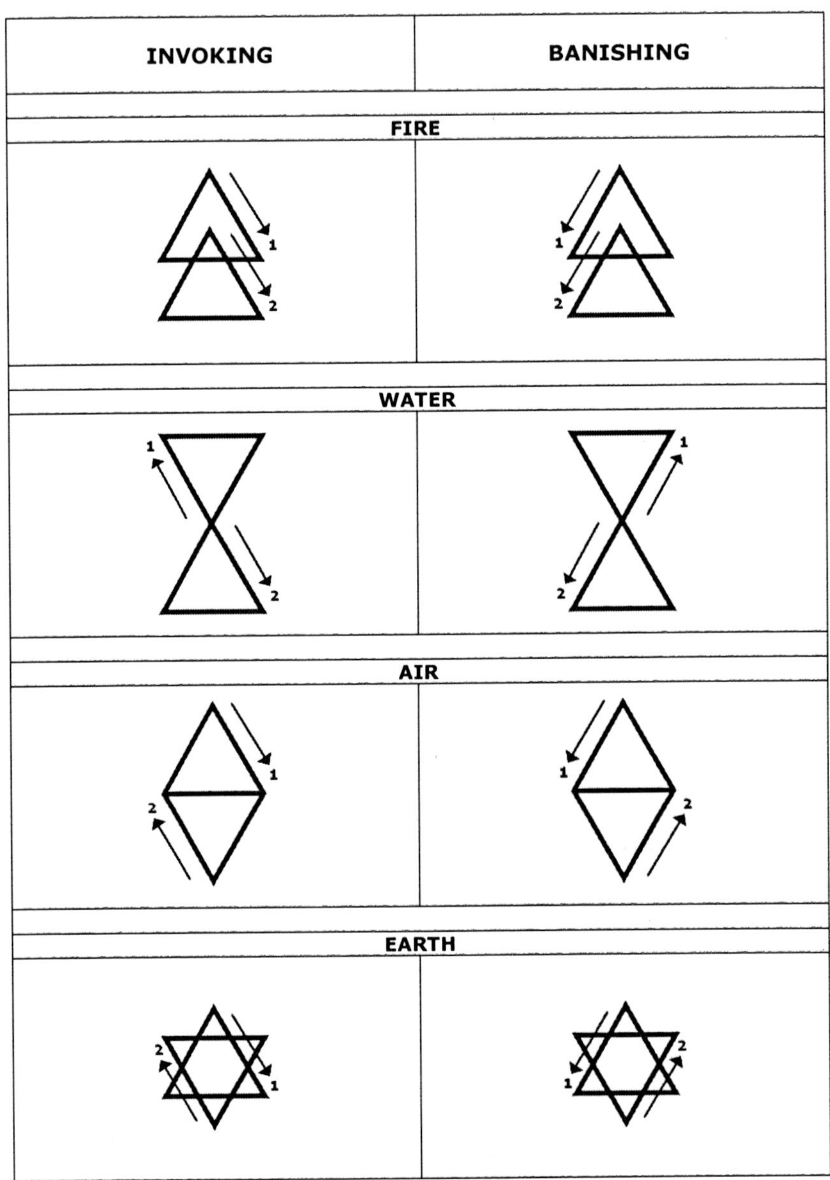

Formula: *ARARITA*

HEXAGRAMS (PLANETARY)

TRADITIONAL		UNICURSAL	
INVOKING	BANISHING	INVOKING	BANISHING
SOL		SOL	

LUNA		LUNA	
 1 2	 1 2		
MARS		MARS	
 1 2	 2 1		
MERCURY		MERCURY	
 1 2	 2 1		
VENUS		VENUS	
 2 1	 1 2		
JUPITER		JUPITER	
 2 1	 1 2		
SATURN		SATURN	
 2 1	 2 1		

FORMULA: *ARARITA*

CHAPTER SEVEN

THE AROUSAL OF APOPHIS
The Dark Night Of The Soul

By now one may be experiencing a slight lethargy concerning the continuation of the work. In fact, one might even find that a mere performance of *Liber Resh* seems totally unthinkable. There is little point in whining about it. It is likely to get much worse.

What, then, is one to do?

Well, the present writer has observed, during this relatively unpleasant phase, that the most agreeable practices seem to be those of an intellectual nature. One found that one's studies in Qabalah proved most productive, as did one's work in divination and ethical questioning. Keeping this in mind, Chapter Seven is composed along these kind of lines.

It must be pointed out though, that one should not cease performing the work set forth in the previous chapters; but to try and embrace these practices wholeheartedly without fear of malfunction. The reason for this is that the Apophis effect is only prolonged by a lack of enthusiasm and concern.

I - Some Encouraging Words

The following essays were written by Crowley and Charles Stansfeld Jones in an attempt to bestow upon the aspirant an understanding of his uniqueness. Each has an uplifting instructive quality, where it is logically reasoned that Thelemic faculty is not only commonsensical, but wholly necessary for the evolution of our species.

DUTY

by

ALEISTER CROWLEY

(A note on the chief rules of practical conduct to be observed by those who accept the Law of Thelema)

"Do what thou wilt shall be the whole of the Law."
"There is no law beyond Do what thou wilt."
"...thou hast no right but to do thy will. Do that and no other shall say nay. For pure will, unassuaged of purpose, delivered from the lust of result, is every way perfect."
"Love is the law, love under will."
"Every man and every woman is a star."

A. YOUR DUTY TO YOURSELF

1. Find yourself to be the centre of your own Universe.

"I am the flame that burns in every heart of man, and in the core of every star."

2. Explore the Nature and Powers of your own Being.

This includes everything which is, or can be for you: and you must accept everything exactly as it is in itself, as one of the factors which go to make up your True Self. This True Self thus ultimately includes all things soever: its discovery is Initiation (the travelling inwards) and as its Nature is to move continually, it must be understood not as static, but as dynamic, not as a Noun but as a Verb.

3. Develop in due harmony and proportion every faculty which you possess.

"Wisdom says: be strong!"
"But exceed! exceed!"
"Be strong, o man, lust, enjoy all things of sense and rapture: fear not that any God shall deny thee for this"

4. Contemplate your own Nature.

Consider every element thereof both separately and in relation to all the rest as to judge accurately the true purpose of the totality of your Being.

5. Find the formula of this purpose, or "True Will", in an expression as simple as possible.

Leave to understand clearly how best to manipulate the energies which you control to obtain the results most favourable to it from its relations with the part of the Universe which you do not yet control.

6. Extend the dominion of your consciousness, and its control of all forces alien to it, to the utmost.

Do this by the ever stronger and more skilful application of your faculties to the finer, clearer, fuller, and more accurate perception, the better understanding, and the more wisely ordered government, of that external Universe.

7. Never permit the thought or will of any other Being to interfere with your own.

Be constantly vigilant to resent, and on the alert to resist, with unvanquishable ardour and vehemence of passion unquenchable, every attempt of any other Being to influence you otherwise than by contributing new facts to your experience of the Universe, or by assisting you to reach a higher synthesis of Truth by the mode of passionate fusion.

8. Do not repress or restrict any true instinct of your Nature; but devote all in perfection to the sole service of your one True Will.

"Be goodly therefore."

"The Word of Sin is Restriction. O man! refuse not thy wife if she will. O lover, if thou wilt, depart. There is no bond that can unite the divided but love: all else is a curse. Accursed! Accursed! be it to the æons. Hell. So with thy all: thou hast no right but to do thy will. Do that and no other shall say nay. For pure will, unassuaged of purpose, delivered from the lust of result, is every way perfect."

"Ye shall gather goods and store of women and Spices; ye shall exceed the nations of the earth is Splendour " pride; but always in the love of me, and so shall ye come to my joy."

9. Rejoice!

"Remember all ye that existence is pure joy; that all the sorrows are but shadows; they pass " are done; but there is that which remains."

"But ye, o my people, rise up and awake! Let the rituals be rightly performed with joy and beauty! .A feast for fire and a feast for water; a feast for life and a greater feast for death! A feast every day in your hearts in the joy of my rapture. A feast every night unto Nuit, and the pleasure of uttermost delight. Aye! feast! rejoice! there is no dread hereafter. There is no dissolution and eternal ecstacy in the kisses of Nu."

"Now rejoice! now come in our splendour and rapture! Come in our passionate peace, & write sweet words for the Kings!"

"Thrill with the joy of life & death! Ah! thy death shall be lovely: whose seeth it shall be glad. Thy death shall be the seal of the promise of our agelong love. Come! lift up thy heart & rejoice!"

"Is God to live in a dog? No! but the highest are of us. They shall rejoice: who sorroweth is not of use. Beauty and strength, leaping laughter and delicious langour, force and fire, are of us."

B. YOUR DUTY TO OTHER INDIVIDUAL MEN AND WOMEN

1. "Love is the law, love under will."

Unite yourself passionately with every other form of consciousness, thus destroying the sense of seperateness from the Whole, and creating a new base-line in the Universe from which to measure it.

2. "As brothers fight ye."

"If he be a king thou canst not hurt him."

To bring out saliently the differences between two points-of-view is useful to both in measuring the position of each in the whole. Combat stimulates the virile or creative energy; and, like love, of which it is one form, excites the mind to an orgasm which enables it to transcend its rational dullness.

3. Abstain from all interferences with other wills.

"Beware lest any force another, King against King!"

(The love and war in the previous injunctions are of the nature of sport, where one respects, and learns from the opponent, but never interferes with him, outside the actual game.) To seek to dominate or influence another is to seek to deform or destroy him; and he is a necessary part of one's own Universe, that is, of one's self.

4. Seek, if you so will, to enlighten another when need arises.

This may be done, always with the strict respect for the attitude of the good sportsman, when he is in distress through failure to understand himself clearly, especially when he specifically demands help; for his darkness may hinder one's perception of his perfection. (Yet also his darkness may serve as a warning, or excite one's interest.) It is also lawful when his ignorance has lead him to interfere with one's will.

All interference is in any case dangerous, and demands the exercise of extreme skill and good judgement, fortified by experience.

To influence another is to leave one's citadel unguarded; and the attempt commonly ends in losing one's own self-supremacy.

5. Worship all!

"Every man and every woman is a star."

"Mercy let be off: damn those who pity."

"We have nothing with the outcast and the unfit: let them die in their misery: For they feel not. Compassion is the vice of kings: stamp down the wretched and the weak: this is the law of the strong: this is our law and the joy of the world. Think not, o king, upon that lie: That Thou Must Die: verily thou shalt not die, but live! Now let it be understood if the body of the King dissolve, he shall remain in pure ecstacy for ever. Nuit Hadit Ra-Hoor-Khuit. The Sun, Strength and Sight, Light these are for the servants of the Star & the Snake."

Each being is, exactly as you are, the sole centre of a Universe in no wise identical with, or even assimilable to, your own. The impersonal Universe of "Nature" is only an abstraction, approximately true, of the factors which it is convenient to regard as common to all. The Universe of another is therefore necessarily unknown to, and unknowable by, you; but it induces currents of energy in yours by determining in part your reactions. Use men and women, therefore, with the absolute respect due to inviolable standards of measurement; verify your own observations by comparison with similar judgements made by them; and, studying the methods which determine their failure or success, acquire for yourself the wit and skill required to cope with your own problems.

C. YOUR DUTY TO MANKIND

1. Establish the Law of Thelema as the sole basis of conduct.

The general welfare of the race being necessary in many respects to your own, that well-being, like your own, principally a function of the intelligent and wise observance of the Law of Thelema, it is of the very first importance to you that every individual should accept frankly that Law, and strictly govern himself in full accordance therewith.

You may regard the establishment of the Law of Thelema as an essential element of your True Will, since, whatever the ultimate nature of that Will, the evident condition of putting it into execution is freedom from external interference.

Governments often exhibit the most deplorable stupidity, however enlightened may be the men who compose and constitute them, or the people whose destinies they direct. It is therefore incumbent on every man and woman to take the proper steps to cause the revisions of all existing statutes on the basis of the Law of Thelema. This Law being a Law of Liberty, the aim of the legislation must be to secure the amplest freedom for each individual in the state, eschewing the presumptuous assumption that any given positive ideal is worthy to be obtained.

"The Word of Sin is Restriction."

The essence of crime is that it restricts the freedom of the individual outraged. (Thus, murder restricts his right to live; robbery, his right to enjoy the fruits of his labour; coining, his right to the guarantee of the State that he shall barter in security; etc.) It is then the common duty to prevent crime by segregating the criminal, and by the threat of reprisals; also, to teach the criminal that his acts, being analyzed, are contrary to his own True Will. (This may often be accomplished b y taking from him the right which he has denied to others; as by outlawing the thief, so that he feels constant anxiety for the safety of his own possessions, removed from the ward of the State.) The rule is quite simple. He who violated any right declares magically that it does not exist; therefore it no longer does so, for him.

Crime being a direct spiritual violation of the Law of Thelema, it should not be tolerated in the community. Those who possess the instinct should be segregated in a settlement to build up a state of their own, so to learn the necessity of themselves imposing and maintaining rules of justice.

All artificial crimes should be abolished. When fantastic restrictions disappear, the greater freedom of the individual will itself teach him to avoid acts which really restrict natural rights. Thus real crime will diminish dramatically.

The administration of the Law should be simplified by training men of uprightness and discretion whose will is to fulfil this function in the community to decide all complaints by the abstract principle of the Law of Thelema, and to award judgement on the basis of the actual restriction caused by the offence.

The ultimate aim is thus to reintegrate conscience, on true scientific principles, as the warden of conduct, the monitor of the people, and the guarantee of the governors.

D. YOUR DUTY TO ALL OTHER BEINGS AND THINGS

1. Apply the Law of Thelema to all problems of fitness, use, and development.

It is a violation of the Law of Thelema to abuse the natural qualities of any animal or object by diverting it from its proper function, as determined by consideration of its history and structure. Thus, to train children to perform mental operations, or to practice tasks, for which they are unfitted, is a crime against nature. Similarly, to build houses of rotten material, to adulterate food, to destroy forests, etc., etc., is to offend.

The Law of Thelema is to be applied unflinchingly to decide every question of conduct. The inherent fitness of any thing for any proposed use should be the sole criterion.

Apparent, and sometimes even real, conflict between interests will frequently arise. Such cases are to be decided by the general value of the contending parties in the scale of Nature. Thus, a tree has a right to its life; but a man being more than a tree, he may cut it down for fuel or shelter when need arises. Even so, let him remember that the Law never fails to avenge infractions: as when wanton deforestation has ruined a climate or a soil, or as when the importation of rabbits for a cheap supply of food has created a plague.

Observe that the violation of the Law of Thelema produces cumulative ills. The drain of the agricultural population to big cities, due chiefly to persuading them to abandon their natural ideals, has not only made the country less tolerable to the peasant, but debauched the town. And the error tends to increase in geometrical progression, until a remedy has become almost inconceivable and the whole structure of society is threatened with ruin.

The wise application based on observation and experience of the Law of Thelema is to work in conscious harmony with Evolution. Experiments in creation, involving variation from existing types, are lawful and necessary. Their value is to be judged by their fertility as bearing witness to their harmony with the course of nature towards perfection.

STEPPING OUT OF THE OLD AEON INTO THE NEW

by

CHARLES STANSFELD JONES

Do what thou wilt shall be the whole of the Law.

As all of you should know, we have entered a New Aeon. A Higher Truth has been given to the World. This truth is waiting in readiness for all those who will consciously accept it, but it has to be realized before it is understood, and day by day those who have accepted this Law, and are trying to live it, realize more and more of its Beauty and Perfection.

The new teaching appears strange at first; and the mind is unable to grasp more than a fragment of what it really means. Only when we are living the Law can that fragment expand into the infinite conception of the whole.

I want you to share with me one little fragment of this great Truth which has been made clear to me this Sun-Day morning: I want you to come with me - if you will - just across the border-line of the Old Aeon and gaze for a moment at the New. Then, if the aspect pleases you, you will stay, or, it may be, you will return for a while, but the road once open and the Path plain, you will always be able to get there again, in the twinkling of an eye, just by readjusting your Inner sight to the Truth.

You know how deeply we have always been impressed with the ideas of Sun-rise and Sun-set, and how our ancient brethren, seeing the Sun disappear at night and rise again in the morning, based all their religious ideas in this one conception of a Dying and Re-arisen God. This is the central idea of the religion of the Old Aeon but we have left it behind us because although it seemed to be based on Nature (and Nature's symbols are always true), yet we have outgrown this idea which is only apparently true in Nature. Since this great Ritual of Sacrifice and Death was conceived and perpetuated, we, through the observation of our men of science, have come to know that it is not the Sun which rises and sets, but the earth on which we live which revolves so that its shadow cuts us off from the sunlight during what we call night. The Sun does not die, as the ancients thought; It is always shining, always radiating Light and Life. Stop for a moment and get a clear conception of this Sun, how He is shining in the early morning, shining at mid-day, shining in the evening, and shining in the night. Have you got this idea clearly in your minds? *You have stepped out of the Old Aeon into the New.*

Now let us consider what has happened. In order to get this mental picture of the ever-shining Sun, what did you do? You identified yourself with the Sun. You stepped out of the consciousness of this planet; and for a moment you had to consider yourself as a Solar Being. Then why step back again? You may have done so involuntarily, because the Light was so great that it seemed as Darkness. But do it again, this time more fully, and let us consider what the changes in our concept of the Universe will be.

The moment we identify ourselves with the Sun, we realize that we have become the source of Light, that we too are now shining gloriously, but we also realize that the Sunlight is no longer for us, for we can no longer see the Sun, any more than in our little old-aeon consciousness we could see ourselves. All around us is perpetual Night, but it is the Starlight of the Body of Our Lady Nuit in which we live and move and have our being. Then, from this height we look back upon the little planet Earth, of which we, a moment ago, were a part, and think of Ourself as shedding our Light upon all those little individuals we have called our brothers and sisters, the slaves that serve. But we do not stop there. Imagine the Sun concentrating His rays for a moment on one tiny spot, the Earth. What happens? It is burnt up, it is consumed, it disappears. But in our Solar Consciousness is Truth, and though we glance for a moment at the little sphere we-have left behind us, and it is no more, yet there is *"that which remains."* What remains? What has happened? We realize that "every man and every woman is a star." We gaze around at our wider heritage, we gaze at the Body of Our Lady Nuit. We are not in darkness; we are much nearer to Her now. What (from the little planet) looked like specks of light, are now blazing like other great Suns, and these are truly our brothers and sisters, whose essential and Starry nature we had never before seen and realized. These are the 'remains' of those we thought we had left behind.

There is plenty of room here, each one travels in His true Path, all is joy.

Now, if you want to step back into the Old Aeon do so. But try and bear in mind that those around you are in reality Suns and Stars, not little shivering slaves. If you are not willing to be a King yourself, still recognize that they have a right to Kingship, even as you have, whenever you wish to accept it. And the moment you desire to do so, you have only to remember this -- Look at things from the point of view of the Sun.

Love is the law, love under will.

II – Tarot Divination

When choosing a suitable Tarot deck, one should keep in mind that one is looking for 78 striking representations of the phenomenal Universe. Therefore, for this reason, the present writer recommends the use of *The Thoth Deck*, which was designed by Aleister Crowley and painted by Lady Frienda Harris. This Tarot pack was six years in the making, and is undoubtedly a most remarkable accomplishment - for every line, symbol, colour and brush stroke was incorporated for a central and underlying purpose.

Here then follows a simple technique in the art and science of Tarot Divination. It is, in fact, one of my own devised methods which is a simplification of a much more complex procedure comprising of 15 cards. The beauty of this system is that one can instantly analyse one's accuracy by checking the card that has fallen on the position designated to the question.

1. Taking the cards in your left hand, hold your right hand over them and make the following invocation whilst imagining a sphere of brilliant white light manifesting directly above your head: "I invoke thee, IAO (pronounced EE-AH-OH), that thou wilt send HRU (pronounced HERU), the great Angel that is set over the operations of this Secret Wisdom, to lay his hand invisibly upon these consecrated cards of art, that thereby we may obtain true knowledge of hidden things, to the glory of thine ineffable Name. Amen."
2. Think of a question and shuffle for as long as feels appropriate. If necessary say the question out loud.
3. Lay the cards out in the following sequence:

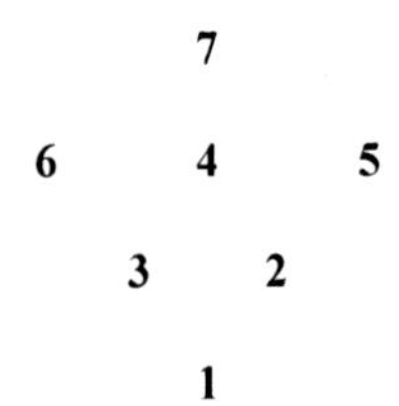

Card 1 is the question and the nature of the problem.
Card 2 is the past and a recent relevant event.
Card 3 shows the Querent in his present situation.
Card 4 gives advice on how to deal with any obstacles.
Card 5 is the path that one's life will naturally follow.
Card 6 suggests an option to the route described in 5.
Card 7 is the outcome or the Karma of the Querent.

4. Make a written interpretation of each card and include it in the magical diary.
5. Refer back to the information at a later date in order to check your accuracy.

It must be remarked that childish superstitions connected with the Tarot should be considered exactly as just that. In fact, the current writer was wholly surprised to discern one evening, on the return from his kitchen with some coffee, that one of his guests - having just previously asked for a general Tarot analysis - appeared to have a considerable amount of anxiety expressed all over his face. When asked what was wrong, he explained that he had just dropped a card whilst shuffling as instructed, and had been previously informed by some halfwit that this meant "irretrievable bad luck". Needless to say one corrected his misunderstanding by dropping all 78 cards at once; proving thus that his informant was an idiot, and that tools of science should never be given to morons.

III - Studies In Qabalah

1. Familiarise yourself fully with the diagram of the Tree of Life (as shown on page 6).
2. Commit the basic correspondences to memory and test your knowledge of such matters.

3. Study the Literal Qabalah and in particular the methods of Gematria and Notariqon.
4. Take pains to discover your own numerical identity (i.e. Find your own holy numbers).
5. Make a careful note of all the information and include it in the magical record.

IV - The Middle Pillar Ritual

This technique originates from a document belonging to the Stella Matutina (a Golden Dawn descendant group) and was later devised into a highly effective working ritual of great potency by Dr. Francis Israel Regardie. Its practice depends entirely on creatively visualising the central spheres of the Tree of Life whilst simultaneously establishing them within the body as points of energy. This is accomplished through the vibration of God-names (as explained in Chapter Five of this manual).

The Middle Pillar Ritual is much more than a mere exercise in creative visualisation, for when performed correctly it not only induces intense bodily relaxation, but releases one's latent and hidden power source which - when rightly accessed - can install within one's self a "deepening ecstasy of being" as Gerald Suster referred to it in his biography of Israel Regardie, *Crowley's Apprentice*.

Regardie believed that the Middle Pillar technique has not only the ability to heal the practitioner working on it, but also to heal a completely separate individual (this is probably why he devoted much of his time and energy into writing an excellent, but grossly under-rated treatise on the matter, *The Art Of True Healing*[52]).

In essence, the forces aroused from a successful performance of the Middle Pillar Ritual are extremely necessary and imperative to the attainment of an effective magical environment. Without this, the magician might be in danger of succumbing to a life-time of pretentious quasi-masonic-style posturing (as is displayed in such groups as the Caliphate O.T.O. today).

Here then follows the technique of the Middle Pillar Ritual. As one starts to work with it regularly, one will begin to understand just why it is useful during the phase of Apophis.

1. Lie down on your back on a hard floor and relax.
2. Breathe slowly, deeply, easily and rhythmically. A cycle of counting 'four' as the breath flows out, as the breath flows in, as the breath is retained and as the breath is expelled is recommended.
3. Now visualise a sphere, roughly four inches in diameter, whirling and glowing with brilliant white light at the crown of the head. Vibrate the name EHEIEH[53]. Do this for ten cycles of breathing.
4. Visualise light descending to form another sphere at the throat, glowing with ultra-violet light. Vibrate the name JEHOVOH ELOHIM[54] for ten cycles.

---

[52] Published by Helios, 1970.

[53] Pronounced Eh-huh-yeh.

Regardie insisted in his *The One Year Manual* that the meaning of these words is really quite irrelevant in the context of this ritual, and that it is, in essence, their "sound value" which is important to us.

[54] Pronounced Yuh-ho-voh Eh-loh-heem.

5. Visualise light descending to form another sphere at the solar plexus, its colour being clear pink rose. Vibrate the name JEHOVOH ELOAH V-DAATH[55] for ten cycles.
6. Visualise light descending to form another sphere at the genitals, this time deep purple in colour. Vibrate the name SHADDAI EL CHAI[56]. Again, ten cycles.
7. Visualise light descending to the feet where another sphere is formed of rich russet brown. Vibrate the name ADONAI HA-ARETZ[57]. Ten cycles.
8. Contemplate the Middle Pillar you have established within yourself. Picture the five central Sephiroth as throbbing with energy on this Middle Pillar of brilliant light which connects them. You will now try to circulate the energies aroused.
9. As you breathe out - still maintaining your rhythm - imagine the energy going down the left side of the body from the head to the feet. As you breathe in, it travels up the right side from the feet to the head. Do this no less than four times (ten is recommended).
10. As you breathe out, imagine the energy pouring down the front of your body from head to feet. As you breathe in, imagine it rippling up the back of your body, from feet to head. Do this four more times.
11. Now throw your attention down to your feet. Imagine the energy rising up through the Middle Pillar to the crown of the head as you breathe in. Then as you breathe out, picture it cascading back down to the feet. Some call this the Fountain Exercise. Do it four or more times.
12. The Exercise of the Interwoven Light: as you breathe in, imagine that a band (or bands) of brilliant white light are weaving around your body, starting from the feet. Like an Egyptian pharaoh, you are being mummified. Continue until your head is finally within a weaving.
13. Express a silent prayer (or prayers) using words which mean something to you.[58]

There are many more applications for the use of this ritual, some of which will be explored a little later.

-oOo-

[55] This is somewhat unnecessarily lengthy. I agree with Regardie's standpoint in *The One Year Manual* that it is perfectly reasonable to replace the original Hebrew name with the much shorter Gnostic name "IAO" (pronounced EE-AA-OH), since I "vibrates extremely well and is much easier to use".

[56] Pronounced Shah-dye-ail-cheye (The Hebrew "ch" is guttural as in the Scottish word "loch").

[57] Pronounced Ah-doh-nye Ha-ah-retz.

[58] This version of the Middle Pillar Ritual was taken from *Crowley's Apprentice* by Gerald Suster, pp. 118-119 (Samuel Weiser, 1990). Suster himself never missed an opportunity to advocate the virtues of Regardie's Middle Pillar technique.

CHAPTER EIGHT

CONTENDING THE RESTRAINT
Outwitting The Destroyer

> The blackness gathers about, so thick, so clinging, so penetrating, so oppressive, that all the other darkness that I have ever conceived would be like bright light beside it... Here abideth terror, and the blind ache of the Soul, and lo! even I, who am the sole light, a spark shut up, stand in the sign of Apophis and Typhon.
>
> (Crowley, A; *The Vision And The Voice*).

I - A Serious Word of Warning

During my own period of Probation, something very curious took place. The subtlety of this occurrence unnerved me so much that one considers it especially appropriate to reproduce the matter fully here.

On November 23rd, 1993, I took the Oath of a Probationer in the highest of elevated spirits. I commenced the work with the greatest intentions, endeavouring to obtain all the necessary documentation.

I had read in *Liber XIII*[59] that after six months certain Probationers were admitted to a ritual known as *XXVIII*. Therefore, being rather zealous towards the work, I strived endlessly to obtain a copy without avail (having no idea what the ceremony was about, but wanting to study it all the same).

By the end of May 1993, the lethargy had certainly begun to set in. Six months on, there was very little improvement having discerned, by this time, that Apophis was the likely cause. On November 24th I made a phone-call to an old friend who had just returned from a visit to the USA. Strangely enough, whilst in the country, he had acquired a number of Thelemic documents which he had been sifting through just prior to my phone-call. Since it was exactly one year and a day since I had sworn the Probationer's Oath, I made a rather impassive enquiry as to whether he was in possession of *Liber Pyramidos*[60]. My friend said that he was and that he would post me a copy of it along with *Ritual XXVIII* which he'd managed to acquire also.

Two days later the documents arrived, and as I opened the package it became rapidly apparent that the Secret Chiefs had seen fit to point out my lack of vigour. For *Ritual XXVIII*, it seemed, was a disciplinary rite designed solely for idle aspirants[61].

Here then follows the transcript of the ceremony in question. Being relatively rare[62], I have reproduced it fully for the scholar and for the Probationer who finds himself trapped in a state of indolence. Be well warned that when a magician strays from his path, the Universe will readjust and imbalance as swiftly and impulsively as possible. Not because it sees fit to single him out, but because he has magically sworn to align himself with the nature and powers of his own being. Nine times out of ten, such readjustments tend to manifest in a most unpleasant and harrowing form.

---

[59] *Vel Gradum Montis Abiegni* (A syllabus of the steps upon the path).

[60] *Ritual DCLXXI* which constitutes the Probationer a Neophyte (see Chapter Ten for the whole transcript of this ceremony).

[61] Needless to say, it was April 1994 before I advanced to the grade of Neophyte.

[62] Today (June 17th 2007) - with present day internet technology - this statement seems particularly laughable. Back in 1993, however, one had to rigorously search for documents that one was never quite sure existed.

RITUAL XXVIII

The Ceremony Of The Seven Holy Kings

(Probationers, who are idle or luxurious, Shall be given a task suitable to their natures. If they refuse the task from laziness or from a feeling that they have more important business (as their Neophyte may judge) then may V.V.V.V.V. 8=3 himself inform them with much deference that they are now fitted for admission to the Mystery of the Seven Holy Kings.)

(The Temple is arranged as shown in the attached illustration [missing in this copy].)

The God of the first Throne hath robes of Deep Blue; second - Violet; 3rd. - Scarlet; 4th. - Orange; 5th. - Green; 6th. - Pale Yellow; 7th. - Bright Blue.

The other officers are the Hegemon, clad in white & masked in white, & in his hand is the Phoenix Wand of the Planets; & the Hiereus, masked in black & robed in black, in his hands are the Scourge & Crook.

The Candidate, well fed & joyful, clad in the Robe of a Probationer, & crowned with laurel, is led by the Hegemon (his Neophyte) into the Hall. Taken to the altar, on which burns a small fire of odorous wood, cedar or sandal or ligri aloes, he is made to kneel thereat, & the Black officer comes forward threatening him with his scourge & saith:

H: Who art Thou?

Hg., for Cand.: I am the Aspirant to the Sacred & Sublime Order of A:.A:. & I seek the aid of Osiris.

(Note: This ceremony can be adapted for working by two officers, the Hegemon assuming each time a different coloured cloak, & the appropriate God-form. But it is most desirable that the full complement of Nine should assist.)

'Hail unto thee, Osiris, triumphant, Lord of Amennta, Lord of Enenet! Hail unto thee, all glorious sun of mercy &justice, upon whose head is the golden crown of light that is invisible to men! Hail unto thee, hail unto thee, sole light in our darkness! Hail unto thee through whom alone (...) may attain unto the Brotherhood Immortal. Deign to guide this aspirant in the straight path, & let him not fall into the way of those who err."

(There is no answer).

"Let us arise, & seek Osiris."

(They come to the First Throne.)

SATURN

"Welcome, welcome, welcome, for thou art chosen, 0 thou that hast aspired to the Brotherhood Immortal. Aspiration is strength, & I give thee of my bounty: Peace & plenty & contentment & good health & length of days. All these hast thou won by virtue of that single aspiration. But beware of the black shadow at my side, for he shall put ice against thy heart; he shall constrict thy whole being; he shall bring thee to sorrow & poverty & premature old age if thou so much as lift thine eyes unto his face. Place therefore thine head upon my knees, that I may put mine hands upon thine head, & bless thee with my blessings."

(He does go.)

"Welcome wast thou, & thou shalt be welcome to my brethren. Pass thou on."

(They reach the Second Throne.)

JUPITER

"Welcome, welcome, welcome, for thou art chosen, 0 thou that hast aspired to the Brotherhood Immortal. Aspiration is strength, & I give thee of my bounty: Authority, & the respect of men, & distinction & praise & veneration. All these hast thou won by virtue of that single aspiration. But beware of the black shadow at my side, for he shall cast thee down, & thou shalt be despised of all men, & thy power shall be broken if thou so much as lift thine eyes unto his face. Place therefore thine head upon my knees, that I may put mine hands upon thine head, & bless thee with my blessings."

(He does so.)

"Welcome wast thou, & thou shalt be welcome to my brethren. Pass thou on."

(They reach the Third Throne.)

MARS

"Welcome, welcome, welcome, for thou art chosen, 0 thou that hast aspired to the Brotherhood Immortal. Aspiration is strength, & I give thee of my bounty: Courage & energy & force; conquest & domimon. Al these hast thou won by virtue of that single aspiration. But beware of the black shadow at my side, for he shall bum thee as with fire, & all that thou hast shalt thou lose. And in thy battles shalt thou be overcome, & thou shalt be broken & ground into dust if thou so much as lift thine eyes unto his face. Place therefore thine head upon my knees, that I may put mine hands upon thine head, & bless thee with my blessings."

(He does so.)

"Welcome wast thou, & thou shalt be welcome to my brethren. Pass thou on."

(They reach the Fourth Throne.)

SUN

"Welcome, welcome, welcome, for thou art chosen, 0 thou that hast aspired to the Brotherhood Immortal. Aspiration is strength, & I give thee of my bounty: Fame & jollity, & a life fair & open; glory & harmony shall be thy servants, & victory shall wait upon thee as an handmaid. All these hast thou won by virtue of that single Aspiration. But beware of the black shadow at my side, for he shall drive thee from the life of men, so that thou hidest in dens & caverns from the light, & thy name shall be lost, & thou shalt suddenly be slain if thou so much as lift thine eyes unto his face. Place therefore thine head upon my knees, that I may put mine hands upon thine head, & bless thee with my blessings."

(He does so.)

"Welcome wast thou, & thou shalt be welcome to my brethren. Pass thou on."

(They reach the Fifth Throne.)

VENUS

"Welcome, welcome, welcome, for thou art chosen, 0 thou that hast aspired to the Brotherhood Immortal. Aspiration is strength, & I give thee of my bounty: Love & beauty & true happiness, ease & abundance. All these hast thou won by virtue of that single aspiration. But beware of the black shadow at my side for he shall destroy love in thee, & all thy beauty shall be blasted, & no word of kindness Shalt thou ever hear again if thou so much as lift thine eyes unto his face. Place therefore thine head upon my knees, that I may put mine hands upon thine head, & bless thee with my blessings."

(He does so.)

"Welcome wast thou, & thou shalt be welcome to my brethren. Pass thou on."

(They reach the Sixth Throne.)

MERCURY

"Welcome, welcome, welcome, for thou art chosen, 0 thou that hast aspired to the Brotherhood Immortal. Aspiration is strength, & I give thee of my bounty Learning, & eloquence, & the power to heal the ills of men. All these hast thou won by virtue of that single aspiration. But beware of the black shadow at my side, for a thief shall come upon thee & despoil thee; & thou shalt have no more knowledge, & with disease thy body shall dissolve away if thou so much as lift thine eyes unto his face. Place therefore thine head upon my knees, that I may put mine hands upon thine head, & bless thee with my blessings.

(He does so.)

"Welcome wast thou, & thou shalt be welcome to my brethren. Pass thou on."

(They reach the Seventh Throne.)

MOON

"Welcome, welcome, welcome, for thou art chosen, 0 thou that hast aspired to the Brotherhood Immortal. Aspiration is strength, & I give thee of my bounty: Purity & clearness of vision, & all the harvest of delight All these hast thou won by virtue of that single aspiration. But beware of the black shadow at my side, for he shall darken thine eyes, & thou shalt waste away, & thou shalt go a-cold, & thou shalt suddenly be slain if thou so much as lift thine eyes unto his face. Place therefore thine head upon my knees, that I may put mine hands upon thine head, & bless thee with my blessings."

(He does so.)

"Welcome wast thou to me, as thou wast welcome to my brethren; thou didst but lift up thine hand in aspiration to the Brotherhood Immortal, & thou hast swept the seven chords of the celestial harp. Pass thou on."

They go back, not to the altar. but to the little door of the temple, within which is a dark dungeon.

The Hegemon thrusts him furiously into this with hand & foot. all crying aloud "Osiris is a black god."

There he must remain for seven hours.

If at any point previous in the ritual he should say "I ask not these blessings; I seek Osiris. What saith yon silent dark one?" or words of similar purport, the Hegemon answers, removing the hoodwink once a for all, "Verily, thou sayest well; know that Osiris Is a black God, & the straight way unto the Sacred & Sublime Order lieth not through the green pastures & beside still waters. But in the Valley of the Shadow of Death, His crook & scourge shall avail thee.

"Take them therefore & fold thine arms upon thy breast; ascend with me the seven steps of the Throne."

The Aspirant then does so, standing between the officers. The Fourth Throne is removed to leave a passage. The Seven Kings rush upon him & belabour him with their weapons; but he makes his way & they fall one by one; the Seventh upon the Second step, the Sixth upon the Third, the Fifth upon the Fourth, the Fourth upon the Fifth, the Third upon the Sixth, the Second upon the Seventh & the First at the Foot of the Throne.

The Aspirant passes over their bodies & takes his seat. They then, each from his place, adore him while the two officers support him on either side, & the Hierophant addresses him:

"Frater............., this day have I symbolically placed thee in the seat of a Brother of the A:.A:. See to it that thy life truly reflect this gradual conquest of the Posers of the Seven, & never forget that thy path is the path of Osiris, - & that Osirjs is a black God."

(They then conduct him from the temple.)

(After the seven hours are past, the Aspirant is rescued by Osiris, the black officer, in the words already given, omitting "Verily thou sayest well" & using a sterner tone for the beginning of the speech. The ceremony then proceeds as before.)

II - Etheric Travel & Astral Projection

So much utter drivel has been written about astral travel that one never ceases to be surprised by the senseless irrelevance that is often generally disclosed to the occult student as esoteric knowledge. Fortunately, however, one occasionally stumbles upon a writer who has properly understood the matter and who clarifies the methodology of such a technique into plain English without resorting to pretentious psycho-babble.

First of all, one must state quite clearly that astral travel and etheric travel are two entirely different things. Although first-hand experience in these fields might procure very similar results, what actually takes place is exceptionally diverse in terms of practical understanding.

Qabalists believe that the etheric body (or "double") is closely associated with the physical body, i.e. that it pertains to the sephira of Malkuth (the material universe in which we each live, breathe and perceive in). One of the main features about travelling in this body is that the experimenter is often able to float out of the physical body (usually via the top of the head), journey across the road to a house he has never visited, for example, and describe its contents and layout in great detail. This can then be subsequently verified by an independent witness. Astral travel, on the other hand, takes place in the subconscious (the sphere of Yesod), the area of the mind which we closely associate with the occurrence of dreams. In this sense then, astral projection can be aptly described as "directly willed day dreaming". Therefore, in order to attain perfect control of this technique, one must first of all gain mastery of one's own imagination.

Many writers will inform you that etheric travel is highly dangerous and that if the connection between the physical body and the etheric double is broken[63], one is very much risking the occurrence of death. Having not yet experienced the joys of etheric travel (nor the event of death for that matter), the present writer is unable to make an educated comment on the nature of such phenomenon. However, one is willing to make an educated guess that jealousy plays an exceedingly large part in the comments of such authors. My advice is simple: open and conclude all experiments with the Lesser Banishing Ritual of the Pentagram, document all experiments fervently, and if you find that you have underwent an experience of etheric travel, congratulate yourself on being one of the very few who can perform this extremely difficult feat of consciousness alteration.

[63] This connection is often described as being a silver cord which connects the physical body with the etheric double.

The Vessel Of Rapid Transcendental EXploration

Many books insist that all astral projection exercises should begin with the practitioner shifting his consciousness into his astral double so that he may then leave his physical body in order to explore the "astral plane". Although there is little doubt that this is an excellent exercise in developing one's will and ability to visualise creatively, one fails to comprehend why it is the only technique utilised (particularly when we know that astral projection is an exemplary method for exploring the subconscious regions of the mind). Certainly it is useful to begin such practices with some form of standard procedure (for how otherwise might we separate astral projection from conventional day dreaming?), however, striving to successfully perform this shift in consciousness can prove to be a most deplorable obstacle for the beginner to overcome. As Aleister Crowley himself wrote:

> Let him imagine his own figure (preferably robed in the proper magical garments, and armed with the proper magical weapons) as enveloping his physical body, or standing near to and in front of him... Let him then transfer the seat of his consciousness to that imagined figure; so that it may seem to him that he is seeing with its eyes, and hearing with its ears.
> This will usually be the great difficulty of the operation.[64]

For this reason then, the present writer has devised a much simpler method which allows one access to the astral plane quickly and effectively without too much strain imposed upon one's concentration. The technique is known as the VORTEX or (**V**essel **Of** **R**apid **T**ranscendental **EX**ploration) and was conceived in order to overcome the problems previously mentioned. Its method of procedure is described as follows:

1. Perform the Lesser Banishing Ritual of the Pentagram and then lie down comfortably (in the foetal or corpse position) with eyes closed.
2. Using the formula ABRAHADABRA, conjure into you imagination a circular tunnel with appears to bend and twist infinitely through the dimensions of space and time (it is likely that each individual will perceive the VORTEX differently since (even on the material plane) it is impossible for two people to experience the same reality. Some may imagine the walls of the VORTEX to be smoke-like, whilst others may see them as being in the category of something more magnetic/electrical. For those finding difficulty in this area, try visualising a "worm hole" or some other kind of anomaly as is often encountered in science fiction television shows).
3. Envision yourself being sucked into the VORTEX by the force of your own will. Imagine that you are travelling along this tunnel at high speed and that it can take you to any given destination conceivable.
4. Conjure up a specific symbol whilst in the VORTEX and see what visions prevail (for example: visualise the sigil of Capricorn, then travel through it until you reach a landscape pertaining to that nature). Also: try exploring some random dimension by leaving the VORTEX at a completely arbitrary point (this can be accomplished by simply penetrating directly through the tunnel's walls).
5. Return from your destination to the VORTEX by simply employing the same technique as set forth in points 2 & 3.

[64] *Magick*, page 458, *Liber O vel Manus et Saggittae* (Guild Publishing, 1989).

6. Before deciding to return to normal consciousness it is imperative for the student to adhere to the following procedure:
   (a) Will the VORTEX to return you to your place of origin (this can be done in a number of ways, namely by "about turning" and retracing your journey back along the tunnel: or by making the Banishing Pentagram of Earth and then visualising your material body lying before you).
   (b) Gradually merge your consciousness with the image of your figure, becoming more slowly aware of your exact material surroundings.
   (c) Awaken giving the sign of Silence/Hoor-pa-kraat.

   (The occasional difficulty in successfully performing any of the above will not result in anything too traumatic, however, one is advised to master the methods of uniting the astral with the material body since long-term risks have never been researched).
7. Perform the Lesser Banishing Ritual of the Pentagram and then document the experience in the magical record.
8. Check the symbolism of your experience in Crowley's *Seven Seven Seven*, and make a careful note of this also.

Some Further Techniques In The Art Of Astral Travel

For those still having difficulty in the field of astral projection, it may be beneficial to explore some of the following innovations:

The Spatial Displacement Method

1. Perform the Lesser Banishing Ritual of the Pentagram.
2. Take a chair and sit yourself down several feet away from where you are going to perform the astral projection.
3. Remaining seated, simply observe the structure and layout of the room in which you are present. Continue doing so for several minutes.
4. Lie down in a comfortable posture with eyes closed.
5. Imagine that there is a humanoid figure present within the room which has just seated itself down on the chair before you.
6. Identify the figure as being your own astral double.
7. Project your consciousness into the figure by simply recalling the images you perceived whilst yourself seated upon the chair.
8. Conjure the VORTEX.

The Stroboscope Method

This technique was first explored by Dr. W. Grey-Walters in *The Living Brain* and later expounded by the artist, writer and traveller Brion Gysin. These two men separately researched the effects of flickering light upon the human mind, which led to the conclusion that under certain frequencies of flicker, some of the physiological barriers between the regions of the brain break down. Using only a light bulb, a 78 rpm gramophone turntable and several metal or card cylinders, Gysin constructed a *Dreamachine* and became fascinated with the results:

> Flicker may prove to be a valid instrument of practical psychology: some people see and others do not. The DREAMACHINE, with its pattern visible to the open eyes, induces people to see. The fluctuating elements of flickered design support the development of autonomous movies, intensely pleasurable and, possibly, instructive to the viewer.
> What is art? What is colour? What is vision? These old questions demand new answers when, in the light of the Dreamachine, one sees all of the ancient and modern abstract art with eyes closed.[65]

Gysin's findings were on a parallel with those of Dr. Grey-Walters, inasmuch that he found the most curious effects seemed to manifest with the machine operating in the region of the Alpha band (8.0 - 13.0 cycles per second). As Simon Dwyer remarked in *Rapid Eye I*:

> The visions hollowed out of the Dreamachine usually start off with a rapid, and quickening, succession of abstract patterns. Often this transit of speeding images is followed by a clear perception of human faces. Humanoid figures and the apparent enactment of highly coloured events, or, as Gysin described them "pseudo events", carried out in time and space.[66]

The present writer, from his own experiences in this intriguing field of study, can only concur with Mr. Dwyer's observations. In fact, at times the effects of strobe-scanning and astral projection seem often so similar in context that making a distinction between the two would be utterly pointless. Certainly more research needs to be carried out in this area, nevertheless, for those experiencing difficulty in the practice of astral projection, the use of a *Dreamachine* would be of undoubted help. These days, however, one need not immerse oneself in the timely construction of such a device. Miniature stroboscopes are available from various electrical shops at a relatively reasonable price, and most of them operate at a maximum flash rate of 10.0 cycles per second (well within the region of the Alpha band). For those interested in exploring this field, the following guidelines should be adhered to rigorously:

1. Commence and conclude all experiments with the Lesser Banishing Ritual of the Pentagram.
2. Switch the rate of the stroboscope to 10 cycles per second (normally "max" on most miniature strobes) and close your eyes (it may be beneficial to have someone else in the room with you during the first few experiments. The reason for this being that strobe lighting has been known to induce epilepsy in a small number of people [one hopes that one's companion isn't susceptible to the negative effects of flickering light also, otherwise one might find oneself becoming involved in the first occurrence of "group epilepsy"]. If at any point you feel nauseous, use common-sense and stop the experiment).
3. Bring the stroboscope to the level of your eyes, then move it closer until a beautiful and complex kaleidoscope of colours manifests on a plane in front of your vision.
4. Simply observe for several minutes until the dream state kicks in. Should you find that nothing whatsoever has occurred beyond the kaleidoscope phenomenon, try reducing the flicker rate slightly until visions appear. Often one finds the image of a tunnel opening up before one. This then is an excellent way to commence the experiment since this aligns strobe-scanning in

[65] *Rapid Eye I*, page 53, *Dreamachine - An Information Montage* by Simon Dwyer (Creation Books, 1995).
[66] ibid, page 54.

accordance with astral projection, and in particular the techniques of conjuring the astral VORTEX.

5. Document all the research diligently in the magical record, then see if one's ability to project astrally (without the machine) has improved (one should find that this *is* the case, since as Gysin observed: "However you look into the Dreamachine, in a short time you will have acquired greater self-knowledge, extended the limits of your vision, brightened your perception of a treasure you may not have known your own"[67].

III - Mind Expansion

The Concise Oxford Dictionary defines the term "reality" as "what is real", "existent" or "underlies appearances". Other than the latter of these definitions such expressions are perhaps a little misleading, for in terms of esoteric understanding, "reality" is something which cannot be defined collectively. For example, although it is perfectly reasonable for two people to concur that "the grass is green", there is absolutely no way that the same two individuals can relate, to each other, their perception of exactly what "green" is. Although both of them might agree that "green" is the colour acquired by mixing "blue" and "yellow" together, both parties have still not defined their comprehension of what is meant by the terminology "colour". How do we know, for example, that person A perceives "green" in the same way that person B perceives "green"? Might A see B's "blue" and call it "green"? In actual fact, both parties are quite at liberty to concur with the statement "the grass is green", merely because they have been programmed from birth to accept such a statement as gospel, thereby assuming that we each perceive the same reality.

To take another example: Magic Johnson, the basketball player, is perfectly free to remark that Lester Piggott, the jockey, is "small". In terms of his height, experience and comprehension, Mr. Johnson's statement is quite correct. On the other hand, Danny DeVito, the Hollywood actor, might meet Mr. Piggott at a film premiere and remark that Mr. Piggott is in fact "quite tall". In other words: Mr. DeVito's comprehension of reality differs considerably to that of Mr. Johnson's. To take matters a ridiculous stage further: Mr. Piggott's wife might consider her husband to be an exceptionally generous lover, whereas the taxman might be more inclined to regard Mr. Piggott as a tight-fisted little wanker. In other words: **the way in which we each perceive reality is fundamentally the result of our upbringing, education and conditioning**. As Robert Anton Wilson wrote in his *Cosmic Trigger I*:

> ...all of our perceptions have gone through myriads of neural processes in the brain before they appear to our consciousness. At the point of conscious recognition, the identified image is organised into a three dimensional hologram which we *project outside ourselves* and call "reality." We are much too modest about our own creativity if we take any of these projections literally. We see the sun "going down" at twilight, but science assures us that nothing of the sort is happening; instead the earth is turning. We perceive an orange as *really* orange, whereas it is actually blue, the orange light being the light bouncing off the real fruit. And, everywhere we look, we imagine solid objects, but science only finds a web of dancing energy.[68]

How then might we improve the perception of our reality matrix? Well, to begin with, proper comprehension of the aforementioned words will already have initiated a

[67] ibid, page 54.
[68] *Cosmic Trigger I*, pp. 28-29 (New Falcon Publications, 1993).

new way of thought for some readers. Once the reader understands fully that the very fabric of reality should be interpreted subjectively, he will be well on his way to the performance of some simple experiments.

In *Liber III vel Jugorum*, Aleister Crowley devised a splendid technique known to most occultists as "the ring method". Its practice is described as follows:

> By some device, such as the changing of thy ring from one hand to the other, create in thyself two personalities, the thoughts of one being within entirely different limits from that of the other, the common ground being the necessities of life... For instance, let A be a man of strong passions, skilled in the Holy Qabalah, a vegetarian, and a keen "reactionary" politician. Let B be a bloodless and ascetic thinker, occupied with business and family cares, an eater of meat, and a keen progressive politician. Let no thought proper to "A" arise when the ring is on the "B" finger, and vice versa.[69]

At this point, the reader might consider the aforementioned practice to be rather dangerous, inasmuch that he may regard such experimentation as a definite path towards developing multiple personality syndrome, schizophrenia, or even advanced hysteria. His assumptions, of course, would be correct; assuming, that is, that the reader is well below average intelligence with a history of mental illness. However, surmising for a moment that this isn't the case, experimentation in this field should result not only in a broadening of the mind's horizons, but in a greater tolerance and understanding for the ideas and concepts of others. As a friend remarked in a short article he wrote: "try watching the 9 O'clock News as a reactionary right-wing Christian fundamentalist and the 10 O'clock News as a liberal left-wing Buddhist."[70]

Although some individuals might feel that this technique will only result in the splitting up (or halving) of the practitioner's current personality, experience with the method should in fact result in both personalities evolving simultaneously beyond the scope of their restricted comprehension. Once this has been realised, one might wish to create a third, fourth, or fifth personality in order to expand further the confines of one's own reality matrix[71].

A final remark might be appropriate here whilst still on the subject of defining reality. Having looked closely at various interpretations and definitions of exactly "what reality is" (from numerous mathematical theories to the proofs dreamt up by the qabalists), the best model that I can come up with is a Vindaloo curry: **there are many ingredients that go towards making a Vindaloo, yet, when one is in the**

[69] *Magick*, page 492 (Guild Publishing, 1989).

[70] *The Source Newsletter*, No. 2 - Autumn Equinox 1996, *Choose Your Masks* by Sagittaria Auria (aka Steve Hind).

[71] In *Cosmic Trigger I*, Robert Anton Wilson remarks that he has as much as 24 different personalities living within him at once; some of which include a Skeptic, a Wizard, a Fool, and a Shaman. Wilson maintains that the Skeptic is the only one who possesses veto power over all the others; a point worth noting for anyone interested in pursuing research in this curious field.

The present writer has also done some research in this area of mind expansion, which has undoubtedly led to much confusion amongst most of his family, friends and colleagues. Terrence Mortimer, for example, is a petulant and outspoken materialist who cares not a damn for 90% of the population. He acts and then thinks later, thus callously attacking anyone who dares to oppose him or stand in his way. George Mortimer, on the other hand, is a diligent and patient occultist who cares intensely about humanity. Each and every action he performs is quickly but conscientiously calculated out.

Initially, each persona seemed to be little more than 50% of the writer's original identity prior to the experiment. However, as time progressed, both personalities seemed to develop into almost two entirely different reality matrices. There is, thankfully, a third personality who holds dominance over the other two. Few indeed are familiar with him, except close friends and, of course, those reading this book.

process of eating the curry, one's experience is limited to the appreciation of the curry as a whole, and not each individual spice that the dish is composed of.

I find reality to be rather similar to a Vindaloo in this respect; for whatever ingredients go towards making up the very nature of the universe, our material interpretation will always be viewed collectively.

-oOo-

CHAPTER NINE

THE COMING OF OSIRIS
The Manifestation Of The Redeemer

> I am the Radiant One, brother of the Radiant Goddess, Osiris the brother of Isis; my son and his mother Isis have saved me from my enemies who would harm me. Bonds are on their arms, their hands and their feet, because of what they have done evilly against me. I am Osiris, the first born of the company of gods, eldest of the gods, heir of my father Geb.
>
> (From *The Egyptian Book Of The Dead*).

A duration of a least eight months should now have elapsed since the swearing of the Oath of a Probationer, and the inescapable lethargy surrounding the phase of Apophis should, at long last, be transforming into a period of gradual productivity. Although there is no reason for the aspirant to have fully mastered any of the aforementioned techniques, one should - at this point - be relatively proficient in the basics of ceremonial magick. So much so, in fact, that the following practices - which require a certain amount of creative imagination - should pose very little problems in terms of effective execution.

I - The Holy Guardian Angel

In his magnum opus, *Magick*, Aleister Crowley wrote:

> ...the Single Supreme Ritual is the attainment of the Knowledge and Conversation of the Holy Guardian Angel. *It is the raising of the complete man in a vertical straight line.*
>
> **Any deviation from this line is black magic. Any other operation is black magic...** If the magician needs to perform any other operation than this, it is only lawful in so far as it is a necessary preliminary to That One Work.[72]

Although, at this stage in one's magical training, there is very little point in striving to fully attain the Knowledge and Conversation of the Holy Guardian Angel (since such an exalted degree of illumination is in fact the primary task of the Adeptus Minor), the present writer does feel that it is of considerable importance for the Probationer to at least grasp a fair comprehension of exactly what the H.G.A. is and the curious way in which it manifests. As Lon Milo DuQuette remarked in *The Magick Of Thelema*:

> It is the acknowledged central truth underlying the world's so-called "great religions" and the key to the understanding of the myths upon which they are founded. The Holy Guardian Angel is the divine object of the devotion of the Bhakti Yogi; Krishna to the Hindu, and Christ to the Christian. No matter by what name or form, the Holy Guardian Angel transfigures the devotee and bestows bliss and the expanded consciousness which is prerequisite to any further spiritual experience or attainment. The Concept of the "H.G.A." is ancient, but the term - Holy Guardian Angel - as it is used in Magick is relatively new.[73]

According to Qabalists, the Holy Guardian Angel resides in the sphere of Tiphareth, or the area of the mind we might call "I", "the centre", "the true self", or "the soul". Therefore, although it is part of our own make-up and configuration, it does appear to

[72] *Magick*, page 294 (Guild Publishing, 1989).
[73] *The Magick Of Thelema*, pp. 133-134 (Samuel Weiser, Inc, 1993).

operate separately from the conscious mind. This then is perhaps the main reason why some have come to call it "God", for when in communication with it the H.G.A. it appears to possess knowledge beyond our comprehension.

Let us analyse all of this more closely. Assume for a moment that you could switch off various parts of your brain just as easily as you could switch off a television set[74]. First of all you would commence with your senses (Malkuth), then you would close down your imagination/subconscious (Yesod); after that you might move on to your thoughts (Hod), before proceeding to switch off your emotions (Netzach). At this point you might like to ask yourself exactly what areas of the mind are still left operational? The answer: "I".

At this stage the more discerning reader might notice an interesting parallel existing between (a) the areas of the mind that we have just hypothetically switched off, and (b) the A.·.A.·.'s magical grade structure; for as you ascend up the grades from Probationer to Adeptus Minor, you are in fact travelling from the sphere of Malkuth (the senses) to Tiphareth (the centre). Thus, one is able to explore each of the regions of the mind in turn, until eventually a magical link has been forged between the temporal and the eternal.

How then might one begin to open up such an Inner Dialogue?

Well, like any other relationship, one might wish to begin by learning the H.G.A.'s name. This can be achieved by employing a number of methods. One such technique that the present writer had success in was by means of Tarot Divination. Here then follows a description of what took place on August 15th, 1995:

5:00 pm - Banished by the Star Ruby ritual.
5:08 pm - Tarot Divination.
Began as usual with the preliminary invocation of IAO.
Asked question: Indicate the number of cards needed to acquire the name of my Holy Guardian Angel?
Shuffled cards and drew one from the pack: The 9 of Wands.
Took this to mean 9 cards.
Replaced card in the pack and re-shuffled.
Asked question: What is the name of my H.G.A.?
Nine cards drawn as follows:

1. 8 of Swords.
2. Atu 0, The Fool.
3. 6 of Disks.
4. 9 of Wands.
5. Ace of Disks.
6. Atu VIII, Adjustment.
7. Prince of Disks.
8. Atu XIII, Death.
9. Atu VI, The Lovers.

Note: Four of the cards are Major Arcana:
Atu 0 = **Aleph** (value 1).
Atu VIII = **Lamed** (value 30).
Atu XIII = **Nun** (value 50).
Atu VI = **Zayin** (value 7).
By Hebrew Gematria **ALNZ** has a value of **88**.

[74] A task that seems immensely difficult to some people judging by the low number of disapproving viewers.

It was at this stage that the present writer noticed something of profound interest, for not only was the value 88 a significant personal number (see Chapter Seven, section III), but it was also 5 away from 93 (the same value of Crowley's own H.G.A. Aiwaz). This became increasingly more significant when I observed that 5 was also the number of cards that still had to be utilised from the divination. Therefore, since 9 cards had been dealt and only 4 of them used, I considered it relatively logical to include the other 5 in the final equation. Noticing that the Hebrew letter Heh (with its English equivalent "E") possessed a Gematric value of 5, I decided to use my intuition at this point and place the letter E between the N and the Z of ALNZ (since Atu XIII and VI were the only two Trumps to come out of the divination together). Thus I assumed ALNEZ as the name of my Holy Guardian Angel. Further ALNZ spelt in full comes to 358, and according to Crowley's book *Seven Seven Seven*, 358 = MShICh (meaning "Messiah") and NChSh (meaning "Serpent").

By following a similar procedure, the student may come to acquire his H.G.A.'s name. Thus, whenever confronted by anxiety, one may make an honest and honourable appeal to the Angel, asking earnestly for the anguish to be lifted. When so doing there is no need to fall to one's knees and grovel like a demented Christian. This is the Aeon of Horus; hence acting like a baboon will only breed contempt. You are, after all, a king, therefore it is only fitting that you should act like one.

Should the aforementioned "naming" technique produce no startling results, make a careful note of what has transpired and return to it at a later date. Note: the present writer needed only to perform this operation once.

II - The Magical Weapons

At this stage in one's magical training, the Probationer might want to construct the four elemental weapons. These consist of The Wand, The Cup, The Dagger, and The Pentacle, and should be thoroughly researched in Crowley's *Magick*, part II. There are many variations of each of these weapons, however this should not deter the student from studying exactly what they each represent magically.

In *Liber A vel Armorum*[75], Crowley gives excellent instruction regarding their assembly, however, the implements described in that document should each be constructed a grade at a time. Therefore, for the time being, one might wish to construct makeshift weapons in order to pursue more elaborate magical workings.

The Wand Of Fire

The Baculum or Wand represents the magician's Will. Its function is to invoke, therefore it should be used in such operations as described in Chapter Six, Section IV. Since it symbolises the magician's Will, it should be perfectly straight and rigid. For this reason, one recommends a bough of hazel approximately 12 inches in length and around half an inch in diameter. Once cut from the tree it should be left to thoroughly dry out before further alterations are made to it. As Aleister Crowley wrote:

> ...the magician will cut the wand from the tree, will strip it of leaves and twigs, will remove the bark. He will trim the ends neatly, and smooth down the knots:- this is the banishing.
>
> He will them rub it with the consecrated oil until it becomes smooth and glistening and golden. He will then wrap it in silk of the appropriate colour:- this is the consecration.

[75] See *Magick*, page 499.

> He will then take it, and imagine that it is that hollow tube in which Prometheus brought down fire from heaven, formulating to himself the passing of the Holy Influence through it. In this and other ways he will perform the initiation, and this being accomplished, he will repeat the whole process in an elaborate ceremony.[76]

This "elaborate ceremony" can be constructed from such techniques as described in Chapter Six, Section IV. For those wishing to perform a *truly* "elaborate" ceremony, Israel Regardie's *The Golden Dawn*[77] gives excellent instruction in this field for each of the four weapons. Alternatively, one might wish to make use of the Middle Pillar technique in order to effectively consecrate these weapons: once you have worked the Middle Pillar all the way through, take the weapon in your hands, cross your legs at the ankles, visualise an egg of energy around you, and as you breathe out using the fourfold rhythmic breath, conjure up an image of the force you wish the object to represent, passing all of the energy you have generated into the weapon in your hands.

Once successfully consecrated, the Wand should be wrapped in red silk.

The Cup Of Water

The Chalice or Cup represents the magician's Understanding. Since it symbolises one's heavenly food, it is often used in acts of purification. The shape of the Cup suggests the female vagina, therefore in this respect the Wand is symbolic of the male phallus.

Any ornamental silver chalice will suffice nicely for this implement, keeping in mind that its size and shape should be in direct proportion to that of all the other weapons. Just as the Wand is wrapped in red silk, so should the Cup be wrapped in blue.

The Dagger Of Air

The Sword or Dagger represents the magician's Reason. Since its shape suggests conflict, it is most often employed in acts of banishing (such as the Lesser Banishing Ritual of the Pentagram).

This implement should possess a double-edged blade, and may be purchased or constructed along similar lines to the one described in *Liber A*. It is wrapped in silk of yellow whenever not in use.

The Pentacle Of Earth

The Disk or Pentacle symbolises the magician's body. For just as the Cup is his heavenly food, so is the Pentacle his earthly food.

Although its functions as a weapon are relatively limited, it does maintain a balance within the temple and, in particular, may be used as a symbol of protection in acts of High Magick. It may also be utilised as a Paten for the Sacrament, but most importantly it serves always as a reminder to the magician of his roots.

It may be fashioned out of wood or wax, and should be inscribed with a symbol to represent the Universe. For those finding difficulty in this area, a green silk sachet of salt or even a humble coin will suffice until such times a Pentacle is required.

---

[76] ibid, pp. 196-197.

[77] Llewellyn Publications.

III – Creating a Magical Environment

The techniques in this section require one to have obtained a relatively thorough knowledge of the correspondence tables in Crowley's book *Seven Seven Seven.* Although the preparations required to perform this kind of ceremony can be, at times, incredibly tedious and lengthy, the results acquired from such preliminaries are indeed justly rewarding.

Say, for example, that you wish to improve your current financial status. One would, first of all, begin by furnishing the place of working in order to suggest the intent of "monetary increase". For example: Jupiter is associated with finance, therefore one would create a particular environment that would reflect the whole concept of Jupiter.

How then does one go about this? Well, according to *Seven Seven Seven*, the number 4 is associated with Jupiter, whilst the Gods who rule the planet are Zeus and Amoun-Ra. The colours associated with Jupiter are violet, blue, purple and azure, whereas its precious stones are Lapis Lazuli and Amethyst. Its relating perfume is Saffron, and its vegetable drug is Cocaine. One might then begin by painting a violet square on the floor (or alter) before enclosing it in a circle of blue. All the relative paraphernalia might then be acquired and arranged in the Temple accordingly. This then would constitute the Preparation of the Place (as described on page 29). One would then proceed to bathe and anoint oneself with Holy Oil before donning the robe and entering the Temple.

It is at this point that the ceremonial work might begin, commencing first of all with a standard banishing ritual, before purifying and consecrating the place of work with the elemental Cup and Wand. At this stage one would affirm the intent verbally, before proceeding to invoke Jupiter by means of one's own ingenium (again the methods of Chapter Six, Section IV might come in relatively useful here).

As Crowley remarked:

> Every Magician must compose his ceremony in such a manner as to produce a dramatic climax. At the moment when the excitement becomes ungovernable, when the whole conscious being of the Magician undergoes a spiritual spasm, at that moment must he utter the supreme adjuration.[78]

At this point the reader might be asking: How in the hell do I compose a ceremony in such a way as to produce a spiritual spasm? The answer: "Inflame thyself in praying."

> The mind must be exalted until it loses consciousness of self. The Magician must be carried forward blindly by a force which, though in him and of him, is by no means that which he in his normal state of consciousness calls I. Just as the poet, the lover, the artist is carried out of himself in a creative frenzy, so must it be for the Magician.[79]

This will prove to be the initial difficulty of the task; however, in order to get the ball initially rolling one might try adapting the following invocation of Thoth whilst in a Temple furnished to Mercury. This invocation, one has found, is an extremely potent one which never fails to produce a result of interest when performed by the present writer.

LIBER ISRAFEL

[78] *Magick*, page 251.
[79] ibid.

SVB FIGVRA LXIV

[This book was formerly called Anubis, and is referred to the 20th key, "The Angel"]

0. The Temple being in darkness, and the Speaker ascended into his place, let him begin by a ritual of the Enterer, as followeth.
1. Procul, O procul este profani.
2. Bahlasti! Ompheda!
3. In the name of the Mighty and Terrible One, I proclaim that I have banished the Shells unto their habitations.
4. I invoke Tahuti, the Lord of Wisdom and of Utterance, the God that cometh forth from the Veil.
5. O Thou! Majesty of Godhead! Wisdom-crowned Tahuti! Lord of the Gates of the Universe! Thee, Thee, I invoke.
O Thou of the Ibis Head! Thee, Thee I invoke.
Thou who wieldest the Wand of Double Power! Thee, Thee I invoke!
Thou who bearest in Thy left hand the Rose and Cross of Light and Life: Thee, Thee, I invoke.
Thou, whose head is as an emerald, and Thy nemmes as the night-sky blue! Thee, Thee I invoke.
Thou whose skin is of flaming orange as though it burned in a furnace! Thee, Thee I invoke.
6. Behold! I am Yesterday, To-Day, and Brother of To-Morrow!
I am born again and again.
Mine is the Unseen Force, whereof the Gods are sprung! Which is as Life unto the Dwellers in the Watch-Towers of the Universe.
I am the Charioteer of the East, Lord of the Past and of the Future.
I see by mine own inward light: Lord of Resurrection; Who cometh forth from the Dusk, and my birth is from the House of Death.
7. O ye two Divine Hawks upon your Pinnacles!
Who keep watch over the Universe!
Ye who company the Bier to the House of Rest!
Who pilot the Ship of Ra advancing onwards to the heights of heaven!
Lord of the Shrine which standeth in the Centre of the Earth!
8. Behold, He is in me, and I in Him!
Mine is the Radiance, wherein Ptah floateth over the firmament!
I travel upon high!
I tread upon the firmament of Nu!
I raise a flashing flame, with the lightning of Mine Eye!
Ever rushing on, in the splendour of the daily glorified Ra: giving my life to the Dwellers of Earth.
9. If I say "Come up upon the mountains!" the Celestial Waters shall flow at my Word.
For I am Ra incarnate!
Khephra created in the Flesh!
I am the Eidolon of my father Tmu, Lord of the City of the Sun!
10. The God who commands is in my mouth!
The God of Wisdom is in my Heart!
My tongue is the Sanctuary of Truth!
And a God sitteth upon my lips.
11. My Word is accomplished every day!
And the desire of my heart realises itself, as that of Ptah when He createth his works!
I am Eternal; therefore all things are as my designs; therefore do all things obey my Word.
12. Therefore do Thou come forth unto me from Thine abode in the Silence:
Unutterable Wisdom! All-Light! All-power!
Thoth! Hermes! Mercury! Odin!
By whatever name I call Thee, Thou art still nameless to Eternity: Come Thou forth, I say, and aid and guard me in this work of Art.
13. Thou, Star of the East, that didst conduct the Magi!
Thou art The Same all-present in Heaven and in Hell!
Thou that vibratest between the Light and the Darkness!
Rising, descending! Changing ever, yet ever The Same!
The Sun is Thy Father!
The Mother the Moon!

The Wind hath borne Thee in its bosom; and Earth hath ever nourished the changeless Godhead of Thy Youth!
14. Come Thou forth, I say, come Thou forth!
And make all Spirits subject unto Me:
So that every Spirit of the Firmament
And of the Ether,
Upon the Earth,
And under the Earth,
On dry land
And in the Water,
Of whirling Air
And of rushing Fire,
And every Spell and Scourge of God the Vast One, may be obedient unto Me!
15. I invoke the Priestess of the Silver Star, Asi the Curved One, by the ritual of Silence.
16. I make open the gate of Bliss; I descend from the Palace of the Stars; I greet you, I embrace you, O children of earth, that are gathered together in the Hall of Darkness.
17. (A pause.)
18. The Speech in the Silence.
The Words against the Son of Night.
The Voice of Tahuti in the Universe in the Presence of the Eternal.
The Formulas of Knowledge.
The Wisdom of Breath.
The Root of Vibration.
The Shaking of the Invisible.
The Rolling Asunder of the Darkness.
The Becoming Visible of Matter.
The Piercing of the Scales of the Crocodile.
The Breaking Forth of the Light!
19. (Follows the Lection.)
20. There is an end of the speech; let the Silence of darkness be broken; let it return into the silence of light.
21. The speaker silently departs; the listeners disperse unto their homes; yea, they disperse unto their homes.

IV - Talismans

According to Crowley:

> ...every object soever is a talisman, for the definition of a talisman is: something upon which an act of will (that is, of Magick) has been performed in order to fit it for a purpose. Repeated acts of will in respect of any object consecrate it without further ado.[80]

In other words, a talisman is a material object which has been endowed with magical qualities in order to accomplish a specific objective. Once constructed and charged, the object is allowed no more than one week to fulfil its function. Should it fail in its task, then the whole operation must be considered unsuccessful.

For those wishing to explore extensively this field of study, one highly recommends Israel Regardie's book *How To Make And Use Talismans* (Thornsons, 1981) where he intelligently investigates the whole concept of these devices and logically sets forth a means by which they can be successfully consecrated by meditative or ceremonial means.

For the present, however, one might wish to try some basic experiments. Having already acquired the ability to invoke and banish various forces of nature (as set forth

[80] ibid, page 249.

in Chapter Six of this manual), one could perhaps set about constructing one of the medieval talismans shown at the end of this chapter. These talismans should be either engraved onto a corresponding metal or drawn on paper with ink of the appropriate colour. Charging can then take place in the temple using such methods as one should, by now, be well familiar with (the Middle Pillar technique is of considerable importance here and may be employed to charge talismans in much the same way as it is used to consecrate elemental weapons). In fact, the present writer has noticed that these talismans are far more effective than those created solely out the imagination, since, due to their age, they have already accumulated a minor charge through their continual use over the centuries. It must be pointed out that all such talismans should remain virgin to the magician. As Crowley remarked:

> It is, of course, very important to keep such an object away from the contact of the profane. It is instinctive not to let another person use one's fishing rod or one's gun. It is not that they could do any harm in a material sense. It is the feeling that one's use of such things has consecrated them to one's self.[81]

He then goes on to display the subtlety of his wit by remarking:

> Of course, the outstanding example of all such talismans is the wife. A wife may be defined as an object specially prepared for taking the stamp of one's creative will. This is an example of a very complicated magical operation, extending over centuries. But, theoretically, it is just an ordinary case of talismanic magick. It is for this reason that so much trouble has been taken to prevent a wife having contact with the profane; or, at least, to try to prevent her.[82]

In some circles it is customary to place considerable significance on the timing of the consecration ritual. In the present writer's opinion this is little more than pure horseshit. What *is* important, however, is the frame of mind in which the magician is in at the time of the ritual. This, and only this, will determine the result of the operation[83].

For those wishing to design their own talismans, one could do far worse than adhere to the following procedures as set forth by Aleister Crowley:

> I used often to make the background of my Talismans of four concentric circles, painting them, the first (inmost) in the King (or Knight) scale, the second in the Queen, the third in the Prince, and the outermost in the Princess scale, of the Sign, Planet, or Element to which I was first devoting it. On this, preferably in the flashing colours, I would paint the appropriate Names and Figures.
>
> Lastly, the Talisman may be surrounded with a band inscribed with a suitable "versicle" chosen from some Holy Book, or devised by the Magician to suit the case.[84]

---

[81] ibid.

[82] ibid.

[83] The exception to this rule might of course be with regards to Lunar talismans; for even today we might still subconsciously associate "increase" at the time of the moon's waxing and "decrease" during the phase in which it wanes.

[84] Crowley, A; *Magick Without Tears*, page 157 (New Falcon Publications, 1991).

SOL (POWER)

LUNA (RECONCILIATION)

MARS (SUCCESS IN WAR)

MERCURY (INFLUENCE)

JUPITER (IMPROVED FINANCE)

VENUS (LOVE)

SATURN (REVENGE)

-oOo-

CHAPTER TEN

CONCLUDING THE PROBATIONARY PERIOD
Neophyte Preliminaries

By now the Probationer should have not only obtained a thorough knowledge of basic Crowleyan Magick but also a certain amount of proficiency in the fundamentals of the Art. By reflecting back on what he has learned to date, he will be more than likely filled with awe at the sheer magnitude of just how far he has come in the space of such a short period. Certainly the horizons of his comprehension will have increased considerably during this time, and one should find that one's perception of the Universe has altered somewhat strangely in accordance with this. At this stage then he should take time to reflect on what has occurred, digesting all of what he has learned so that these changes can implement themselves naturally. Once this has been achieved he should start to consider how he is going to prepare himself for the next evolutionary leap, should he desire to proceed with such a thing.

In a short work entitled *Liber XIII*[85], Crowley informs us that the Probationer should pass Ritual DCLXXI (or *Liber Pyramidos*) in order to become a fully fledged Neophyte. However, the present writer found this "Ritual of Self Initiation" to be rather cumbersome thereby coming to the conclusion, for the first time ever, that this was a work which certainly didn't display Crowley at his best. Having discussed the matter with a number of authorities, one found that one was not alone in one's feelings regarding the situation. Having been advised quite categorically to omit the ritual from the curriculum, the present writer set about advancing to the grade of Neophyte in much the same way as he advanced to the grade of Probationer. Looking back in retrospect, one can safely say that the choice was a wise one, and one would certainly urge the reader to take heed of this advice.

However, for those zealous enough to perform the awkward little ceremony I have reproduced it here, in full, as follows. Certainly one could make better use of the time spent learning this ritual, by going over all that one has learned to date. Ten months is certainly not nearly enough time to fully digest all the techniques that have been presented in this manual.

DCLXXI

LIBER PYRAMIDOS

A Ritual of Self Initiation
Based Upon the Formula of the Neophyte

By

Aleister Crowley

BUILDING OF THE PYRAMID

[85] From *The Equinox*, volume III, page 3.

The Magus with Wand. On the Altar are incense, Fire, Bread, Wine, the Chain, the Scourge, the Dagger & the Oil. In his left hand the Bell he taketh.

Hail! Asi! Hail, Hoor-Apep! Let
The Silence speech beget!

Two strokes on Bell. Banishing Spiral Dance

The Words against the Sons of Night
Tahuti speaketh in the Light.
Knowledge & Power, twin warriors, shake
The Invisible; they roll asunder
The Darkness; matter shines, a snake.
Sebek is smitten by the thunder -
The Light breaks forth from Under.

He goes to the West, in the centre of the base of the Triangle of Thoth (Aleph), Asi (Mem), & Hoor (Shin).

O Thou, the Apex of the Plane,
With Ibis head & Phoenix Wand
And Wings of Night! Whose serpents strain
Thou in the Light & in the Night
Art One, above their moving might!

He lays the Wand, etc., on the Altar, uses the Scourge on his buttocks, cuts a cross with the Daggar upon his breast & tightens the Chain of the Bell about his forehead, saying:

The Lustral Water! Smite thy flood
Through me - lymph, marrow & blood!

Anointing the Wounds, say:

The Fire Informing! Let the Oil
Balance, assain, assoil!

The Invoking Spiral Dance (widdershins).

So Life takes Fire from Death, & runs
Whirling amid the Suns.
Hail, Asi! Pace the Path, bind on
The girdle of the Starry One!

Sign of the Enterer:

Homage to Thee, Lord of the Word!

Sign of Silence:

Lord of Silence, Homage to Thee!

Repeat both Signs:

Lord we adore Thee, still & stirred Beyond Infinity.

The Secret Word: MTzThBTzM.

For from the Silence of the Wand
Unto the Speaking of the Sword,
And back again to the Beyond,

This is the toil & the Reward.
This is the Path of HVAÄHo!
This is the Path of IAO.

Bell.

Hail Asi! Hail, thou Wanded Wheel!
Alpha & Delta kissed & came
For five that feed the Flame.

Bell.

Hail, Hoor-Apep! thou Sword of Steel!
Alpha & Delta & Epsilon
Met in the Shadow of the Pylon
And in Iota did proclaim
That tenfold core & crown of flame.
Hail, Hoor-Apep! Unspoken Name!

Thus is the Great Pyramid duly builded.

INITIATION FOLLOWETH

THE FIRST PYLON

The Candidate still bound and hoodwinked.

I know not who I am; I know not whence I came;
I know not whither I go; I seek - but What I do not know!

I am blind & bound; but I have heard one cry
Ring through Eternity; Arise & follow me!

Asar Un-nefer! I invoke
The four-fold Horror of the Smoke.
Unloose the Pit! by the dread Word -
Of Power - that Set-Typhon hath heard –

SAZAZ SAZAZ ANDATSAN SAZAZ

(Pronounce this backwards. But it is very dangerous. It opens up the Gates of Hell.)

The Fear of Darkness & of Death
The Fear of Water & of Fire
The Fear of the Chasm & the Chain
The Fear of Hell & the dead Breath.
The Fear of Him, the Demon dire
That on the Threshold of the Inane
Stands with his Dragon Fear to slay
The Pilgrim of the Way.
Thus I pass by with Force & Care,
Advance with Fortitude & Wit,
In the straight Path, or else Their Snare
Were surely Infinite.

PASSING OF THE SECOND PYLON

Suit action to words.

Asar! who clutches at my throat?
Who pins me down? Who stabs my heart?
I am unfit to pass within this Pylon of the Hall of Maat.

Rubric as above.

The Lustral Water! Let thy flood
Cleanse me - lymph, marrow, & blood!
The Scourge, the Dagger & the Chain
Purge body, breast & brain!
The Fire Informing! Let the Oil
Balance, assain, assoil!

Still in corpse-position.

For I am come with all this pain,
To ask admission to the Shrine.
I do not know why - I ask in vain -
Unless it be that I am Thine.

I am Mentu his truth-telling brother,
Who was Master of Thebes from my birth:
O heart of me! heart of my mother!
O heart that I had upon the earth!
Stand not thou up against me as a witness!
Oppose me not, judge, in my quest!
Accuse me not now of unfitness
Before the great God, the dread Lord of the West!

Speak fair words for (candidates name).
May he flourish
In the place of the weighing of hearts
By the marsh of the dead, where the crocodiles nourish
Their lives on the lost, where the Serpent upstarts.
- For though I be joined to the Earth,
In the Innermost Shrine of Heaven am I.
I was Master of Thebes from my birth;
Shall I die like a dog? Thou shalt not let me die,
But my Khu that the teeth of the crocodiles sever
Shall be mighty in heaven for ever & ever!
Yea! but I am a fool, a flutterer!
I am under the Shadow of the Wings!

Refrain after each accusation.

I am a liar & a sorcerer
I am so fickle that I scorn the bridle.
I am unchaste, voluptuous and idle.
I am a bully & a tyrant crass,
I am as dull & as stubborn as an ass;
I am untrusty, cruel & insane,
I am a fool & frivolous & vain.
I am a weakling & a coward; I cringe,
I am a catamite & cunnilinge.
I am a glutton, a besotted wight;
I am a satyr & a sodomite.
I am as changeful & selfish as the Sea.
I am a thing of vice & vanity.
I am most violent & I vacillate,
I am a blind man & emasculate.

I am a raging fire of wrath - no wiser!
I am a blackguard, spendthrift & a miser.
I am obscene & devious & null.
I am ungenerous & base & dull.
I am not marked with the white Flame of Breath.
I am a Traitor! - die the traitor's death!

This last raises Candidate erect. Invoking Spiral Dance. Rubric as before.

I am under the Shadow of the Wings.
Now let me pace the Path, bind on
The girdle of the Starry One!
Asar! k.t.l.

In Northwest. See Horus.

Soul-mastering Terror is thy name!
Lord of the Gods! Dread Lord of Hell!
I am come. I fear Thee not. Thy flame
Is mine to weave my maiden spell!
I know Thee, & I pass Thee by.
For more that Thou am I!
Asar! k.t.l. (*Rubric as usual*)

In Southwest. See Isis.

Sorrow that eateth up the soul!
Dam of the Gods! The blue sky's Queen!
This is Thy Name. I come. Control
And Pass! I know Thee, Lady of Teeu!
I know Thee & I pass Thee by.
For more than Thou am I!
Asar! k.t.l. (*Rubric as usual*)

In East. See Thoth. Silence

Asar! k.t.l. (*Rubric as usual*)

See Nature.

I will not look upon thee more,
For Fatal is Thy Name. Begone!
False Phantom, thou shalt pass before
The frowning forehead of the Sun.
I know Thee; & I pass thee by.
For more that Thou am I.

Formulating Hexagram.

Now Witness Ye upon the Earth,
Spirit & Water & Red Blood!
Witness Above, bright Babe of Birth,
Spirit, & Father - that are God!

As babe in egg, being born.

For Silence duly is begot
And Darkness duly brought to bed;
The Shroud is figured in my Thought,
The Inmost Light is on my head.

Unbind.

Attack! I eat up the strong lions. I!
Fear is on Seb, on them that dwell therein,

Sign of the Enterer.

Behold the radiant Vigour of the Lord!

Sign of Silence.

Defense! I close the mouth of Sebek, ply
My fear on Nile, Asar that held not in!
Behold my radiant Peace, ye things abhorred
For see! The Gods have loosed mine hands:
Asar unfettered stands.
Hail, Asi, hail! Hoor-Apep cries -
Now I the Son of Man arise
And followÄdead where Asar lies!

Lies down in Sign of Hanged Man.

I guild my left foot with the Light.
I guild my Phallus with the Light.
I guild my right knee with the Light.
I guild my right foot with the Light.
I guild my left knee with the Light.
I guild my Phallus with the Light.
I guild my elbow with the Light.
I guild my navel with the Light.
I guild my heart wedge with the Light.
I guild my black throat with the Light.
I guild my forehead with the Light.
I guild my Phallus with the Light.

Rising in Sign Mulier.

Asar Un-nefer! I am Thine,
Waiting Thy Glory in the shrine.
Thy bride, Thy virgin! Ah, my Lord.
Smite through the Spirit with Thy Sword!
Asar Un-nefer! rise in me,
The chosen catamite of Thee!
Come! Ah, come now! I wait, I wait,
Patient - impatient slave of Fate,
Bought by Thy glance - Come now! come now!
Touch & inform this burning brow.
Asar Un-nefer! in the shrine,
Make Thou me wholly Thine!

Remove hoodwink.

I am Asar - worthy alone
To sit upon the Double Throne.
Attack is mine & mine defence.
And these are one. Arise, go hence!
For I am Master of my Fate,
Wholly Initiate.

The Secret Word: MTzThBTzM

The Words are spoken duly.
The deeds are duly done.
My soul is risen newly
To greet the risen Sun.

Bell accordingly.

One! Four! Five! Ten! Hail!
One! Four! Five! Ten! All Hail!

I give the sign that rends the Veil.
The sign that closes up the Veil.

SEALING OF THE PYRAMID

Proceed as in the Building, unto the word "Suns."

Banishing Spiral Dance.

Now let mine hands unloose the sweet
And shining girdle of Nuit!

The Adorations & the Word. Then at the Altar.

Behold! the Perfect One hath said
Tried & found pure, a golden spoil.
These are my body's elements.

Act accordingly.

Incense & Wine & Fire & Bread
These I consume, true Sacraments,
For the Perfection of the Oil
For I am clothed about with flesh
And I am the Eternal Spirit.
I am the Lord that riseth fresh
From Death, whose glory I inherit
Since I partake with him. I am
The Manifestor of the Unseen.
Without me all the land of Khem
Is as if it had not been.

Proceed as in Building to end.

Hail, Hoor! Hail, Asi! Hail, Tahuti! Hail,
Asar Un-nefer! through the rended Veil.
I am Thyself, with all Thy brilliance decked -
Khabs-Am-Pekht.

-oOo-

AFTERWORD

As previously outlined, one of my main intentions for composing this little work was to "revise and update a very pure, noble and beautiful system that is all too often ignored by so-called magicians working within a Thelemic context." This I believe I have done as best as any individual could do considering the vastness of the task at hand. However, in reality, the revision of the A∴A∴'s system has only just begun for, as I'm sure the reader will by now agree, the syllabus, as set forth by Crowley and Jones in 1907, has become somewhat dated; not so much with respects to its training, but more concerning its structure.

In his writings, Aleister Crowley never missed an opportunity to accentuate the fact that the A∴A∴'s method was one of science. Magick, then, like all other branches of study, is an evolving process which requires perpetual amendment and reassessment at key points in human development.

Today, after the 50th anniversary of Crowley's death, this task seems more important than ever. Although Thelema is gifted with many fine writers and authorities, there is all too often an unfortunate tendency for Thelemites to bicker and argue about trivial and altogether pointless matters of very little consequence; thus restricting not only collaboration, but the evolutionary process of this fine system.

The Caliphate O.T.O. is one such group which should be held responsible for lack of innovation. Its reluctance to evolve has given rise to a number of different movements of which Chaos Magic is but one. This peculiar branch of "Magic" was founded by one Peter J. Carroll in the late-70s for no other reason than what appears to be utter frustration on Carroll's behalf. Having studied the ideas and origins of Chaos Magic, one can certainly sympathise with just how frustrated Mr. Carroll must have felt at the time. Indeed, to begin with, Chaos Magic gave rise to some splendid innovation, combining the techniques of Austin Osman Spare with those of Aleister Crowley.

Alas, however, this creative frenzy was not to continue. Today, Chaos Magic is not only and adolescent mess, but completely flawed in its principle comprehension of basic magical theory[86]. As the writer James Barter pointed out in the *Skoob Esoterica Anthology*:

> **Chaos** is defined as both "a formless void from which the cosmos was evolved", and "utter confusion." **Magic**, by contrast, is the art of causing change to occur in an ordered (i.e., structured) way in existing phenomena. It is caused by a relationship existing between things: at its most basic, a thing to be changed, or *first matter*; and the willed intention to change a thing to a desired end, or *final result*. This relation cannot operate in a void where no thing exists, nor can it exist in a condition of utter confusion, because no points of reference are then involved. The phrase itself is therefore a contradiction in terms: there can be no such thing as a Magic of Chaos, or "Chaos Magic". Q.E.D.[87]

[86] Rereading this as I revise this book today (July 13th 2007) it is probably an unfair statement given how far Chaos Magic has come since initially writing this book. My opinions have changed considerably over the years, however, it seems appropriate to leave my comments as they were written back in 1998.

[87] From an article entitled *Eight Arrows Going Nowhere* by James A. Barter, page 162 of the *Skoob Esoterica Anthology* (Skoob Books Publishing Ltd, 1995).

The task of revising the whole A.·.A.·. syllabus is a colossal one, nevertheless it is one that is necessary and considerably worthwhile. One need only peruse over the Tasks of the Grades to determine that the undertaking is essential. Furthermore, magical techniques have evolved considerably since Crowley's death, and although it is not a case of replacing some of Uncle Aleister's prescribed methods, it is certainly necessary to update and add new ones to the curriculum. To do so is well beyond the scope of one individual's work; what *is* required is a collaboration from Thelemites of all persuasions and aspects of life. Thus the A.·.A.·. system can avoid become as obsolete as the system of the O.T.O.

By setting aside our differences in this manner and concentrating on our similarities, Thelema can once again be at the forefront of occultism at the beginning of the 21st century.

George T. Mortimer.
May 1998 e.v.

-oOo-

APPENDIX A

X-RAYS ON EX-PROBATIONERS

RATS leave sinking ships; but you cannot be sure that a ship will sink because you see a rat running away from it. The captain may have given orders about it.

Persecution is like Keating's Powder. It does not injure the most delicate skin, but it removes all vermin.

"Mine own familiar friend whom I trusted lifted up his heel against me" - and then I saw it was the hoof of an ass.

PERDURABO.

Taken from *The Equinox*,
Volume V, page 142.

APPENDIX B

THE TASK OF A NEOPHYTE

0. Let any Probationer who has accomplished his task to the satisfaction of the A.·.A.·.be instructed in the proper course of procedure: which is:- Let him read through this note of his office, and sign it, paying the sum of One Guinea for *Liber VII* which will be given him on his initiation, and One Guinea for this Portfolio of Class D publications, B - G. Let him obtain the robe of a Neophyte, and entrust the same to the care of his Neophyte. He shall choose a new motto with deep forethought and intense solemnity, as expressing the clearer consciousness of his Aspiration which the year's Probation has given him. Let him make an appointment with his Neophyte at the pleasure of the latter for the ceremony of Initiation.

1. The Neophyte shall not proceed to the grade of Zelator in less than eight months; but shall hold himself free for four days for advancement at the end of that period.

2. He shall pass the four tests called the Powers of the Sphinx.

3. He shall apply himself to understand the nature of his Initiation.

4. He shall commit to memory a chapter of *Liber VII*; and furthermore, he shall study and practice a chapter of *Liber O* in all its branches: also he shall begin to study *Liber H* and some commonly accepted method of divination. He will further be examined in his power of Journeying in the Spirit Vision.

5. Beside all this, he shall perform any tasks that his Zelator in the name of the A.·.A.·.and by its authority may see fit to lay upon him. Let him be mindful that the word Neophyte is no idle term, but that in many a subtle way the new nature will stir within him, when he knoweth it not.

6. When the sun shall next enter the sign 240 degrees to that under which he hath been received, his advancement may be granted unto him. He shall keep himself free from all other engagements for four whole days from that date.

7. He may at any moment withdraw from his association with the A.·.A.·., simply notifying the Zelator who introduced him.

8. He shall everywhere proclaim openly his connection with the A.·.A.·.and speak of It and Its principles (even so little as he understandeth) for that mystery is the enemy of Truth. Furthermore, he shall construct the magic Pentacle, according to the instruction in *Liber A*. One month before the completion of his eight months, he shall deliver a copy of his Record to his Zelator, pass the necessary tests, and repeat to him his chosen chapter of *Liber VII.*

9. He shall in every way fortify his body according to the advice of his Zelator, for that the ordeal of advancement is no light one.

10. Thus and not otherwise may he obtain the great reward: YEA, MAY HE OBTAIN THE GREAT REWARD!

APPENDIX C

A.·.A.·.

The Oath of a Neophye

I, ... [old motto], being of sound mind and body, and prepared, on this .... day of .... (Anno ...... Sol in ......° of ....) do hereby resolve: in the presence of ...., a Zelator of the A.·.A.·.: To prosecute the Great Work: which is, to obtain control of the nature and powers of my own being.

Furthermore, I promise to observe zeal in service to the Probationers under me, and deny myself utterly on their behalf.

May the A.·.A.·. crown the work, lend me of Its wisdom in the work, enable me to understand the work!

Reverence, duty, sympathy, devotion, assiduity, trust do I bring to the A.·.A.·. and in eight months from this date may I be admitted to the knowledge and conversation of the A.·.A.·.!

Witness my hand [*old motto*] ....

New Motto ....

The writer can be contacted through the email address:
mort@media-underground.net

CPSIA information can be obtained at www.ICGtesting.com
Printed in the USA
LVOW10s2125290414

383803LV00002B/17/P